Radiometry and the
Detection of
Optical Radiation

Radiometry and the Detection of Optical Radiation

ROBERT W. BOYD
Institute of Optics
University of Rochester

John Wiley & Sons
New York / Chichester / Brisbane / Toronto / Singapore

Library of Congress Cataloging in Publication Data:

Boyd, Robert W., 1948–
    Radiometry and the detection of optical radiation.

    (Wiley series in pure and applied optics, ISSN 0277-
2493)
    Includes index.
    1. Radiation—Measurement. 2. Optical detectors.
I. Title. II. Series. [DNLM: 1. Radiation, Non-
Ionizing. 2. Radiometry. TA 1522 B78Or]

QC475.B67 1983   535'.028'7   82-2-0222
ISBN 0-471-86188-X

Printed in the United States of America

10 9 8 7 6 5 4

Preface

Much of what is known about the physical universe results from optical measurements, and thus our knowledge of physical laws is often limited by the accuracy to which such measurements can be conducted. The limitations imposed by measurement accuracy are often of a rather fundamental nature. For example, quantum and thermal fluctuations of a radiation field limit the accuracy with which the power carried by the field can be determined. Similarly, thermodynamic considerations limit the total power that can be transmitted through an optical system from a given source, and the total power that can be emitted by a source in thermal equilibrium is likewise constrained. Thus an understanding of the basic principles governing the generation, transfer, and detection of optical radiation is needed in order to properly design and execute many laboratory measurements.

This book discusses some of the fundamental aspects of the generation of radiation by thermal sources, of the transfer of radiation through optical systems, and of the detection of optical and infrared radiation. It is intended to serve as a textbook for first-year graduate students. The material covered here comprises the syllabus of a course in radiometry and detection that has for many years been taught at the Institute of Optics of the University of Rochester. Although a number of extremely good monographs treating radiometry and detection are in existence, there has been a lack of books suitable as texts at the graduate level. This book is intended to present a unified treatment of radiometry and detection at a level that will allow the student to progress to the more advanced works on these subjects, while also presenting enough practical information to serve the student's immediate needs.

The material presented here has been organized as follows: The first chapter presents a brief review of the classical theory of electromagnetism based on Maxwell's equations. The basic properties of the electromagnetic field that are needed for the ensuing development are derived here. The mksa system of units is used here and, with a few exceptions, throughout the text. Chapters 2 and 4 present a discussion of classical radiometry, that is, the radiometry of essentially incoherent sources. [The study of the radiometric properties of partially coherent sources still constitutes an active research area and thus is not discussed in the text. A review of recent research in this area has been given by E. Wolf, *J. Opt. Soc. Am.*, **68**, 6 (1978).] The theory of blackbody radiation is presented in Chapter 3. It proves useful to treat blackbody radiation midway through the discussion of radiometry, so that radiometric notation can be used in the description of the blackbody field, yet the thermodynamic properties of blackbody radiation can be utilized in the discussion of radiometry. Applications of blackbody radiation theory are presented in Chapter 5, and photometry and vision are discussed in Chapter 6.

The detection of radiation is discussed in the second half of the book. Chapter 7 presents a phenomenological discussion of radiation detectors and introduces the terminology conventionally used to describe their performance. Chapter 8 discusses noise, which is present in all detection systems and limits a system's ability to detect weak optical signals. Detailed descriptions of several of the most common types of detectors are presented in Chapters 9 through 13. The last chapter discusses the fundamental limitations to the sensitivity of radiation detectors set by quantum and thermal fluctuations of the radiation field itself.

I wish to thank my colleagues at the Institute of Optics for the assistance and encouragement they have provided during the preparation of this work. I wish particularly to thank J. C. Dainty, G. M. Morris, D. C. Sinclair, and K. J. Teegarden for useful discussions concerning the subject matter of this book, and P. W. Baumeister for making available to me his knowledge and experience in the teaching of this subject. I also wish to thank C. H. Townes for discussions regarding optical radiation and its interaction with matter. Finally, I acknowledge the encouragement provided by my family.

Robert W. Boyd

Rochester, New York
December 1982

Contents

Radiometry and the Detection of Optical Radiation

1

Electromagnetic Radiation

This chapter presents a brief account of the theory of electromagnetism based on Maxwell's equations. These equations govern the behavior of the electromagnetic fields in either free space or in a material medium; they describe both how the fields are generated by free charges and currents and how the fields propagate in the absence of such sources. These equations thus represent the most fundamental non-quantum-mechanical theory of electromagnetism. Certain inherently quantum mechanical features of electromagnetism, such as the quantization of field energy leading to the concept of photons, lie outside the domain of Maxwell's theory and will be introduced later as the need arises. The treatment presented here uses the rationalized mksa system of units, which is best suited to the rather practical nature of the subjects treated in this text.

1.1 MAXWELL'S EQUATIONS

Maxwell's equations have the form

$$\nabla \times \mathbf{E} = -\frac{\partial \mathbf{B}}{\partial t}, \tag{1.1a}$$

$$\nabla \times \mathbf{H} = \frac{\partial \mathbf{D}}{\partial t} + \mathbf{J}, \tag{1.1b}$$

$$\nabla \cdot \mathbf{B} = 0, \tag{1.1c}$$

$$\nabla \cdot \mathbf{D} = \rho. \tag{1.1d}$$

The quantities that appear in these equations and their mksa units are defined as follows:

$$\mathbf{E} = \text{electric field} \qquad (\text{V}/\text{m})$$

$$\mathbf{D} = \text{electric displacement} \qquad (\text{C}/\text{m}^2)$$

$$\mathbf{B} = \text{magnetic induction} \qquad (\text{Wb}/\text{m}^2),$$

$$\mathbf{H} = \text{magnetic intensity} \qquad (\text{A}/\text{m}),$$

$$\rho = \text{free charge density} \qquad (\text{C}/\text{m}^3),$$

$$\mathbf{J} = \text{free current density} \qquad (\text{A}/\text{m}^2).$$

Maxwell's equations implicitly assume the conservation of charge, which can be demonstrated by deriving the equation of charge continuity from them. By taking the divergence of Eq. (1.1b) and using the identity

$$\nabla \cdot (\nabla \times \mathbf{A}) = 0,$$

which is true for any $\mathbf{A}$, one obtains the equation

$$\nabla \cdot \mathbf{J} + \frac{\partial}{\partial t}(\nabla \cdot \mathbf{D}) = 0.$$

Through the use of the Maxwell equation (1.1d), this relation becomes

$$\nabla \cdot \mathbf{J} + \frac{\partial \rho}{\partial t} = 0, \qquad (1.2)$$

which expresses charge conservation by showing that the free charge density ρ can decrease only as a result of a divergent current density $\mathbf{J}$.

The field vectors that appear in Maxwell's equations are further related by conditions that depend on the properties of the medium containing the electromagnetic field. In many cases, particularly when the field strengths $\mathbf{E}$ and $\mathbf{H}$ are not too large, it is possible to assume that a linear relationship exists between several of these quantities. These constitutive relations are given as follows

$$\mathbf{D} = \varepsilon\mathbf{E}, \qquad (1.3a)$$

$$\mathbf{B} = \mu\mathbf{H}, \qquad (1.3b)$$

$$\mathbf{J} = \sigma\mathbf{E}, \qquad (1.3c)$$

where, for convenience, we have considered the case of an isotropic medium so that the proportionality constants ε, μ, σ can be taken as scalars.

The quantity ε is called the dielectric permittivity and has units F/m, where the farad is a unit of capacitance and is equivalently expressed as F = C/V. For fields in free space (i.e., in a vacuum), the permittivity has the value $\varepsilon_0 = 8.85 \times 10^{-12}$ F/m; in a material medium, however, the dielectric constant $\kappa_e = \varepsilon/\varepsilon_0$ will in general not equal unity. This effect occurs because in the presence of an electric field $\mathbf{E}$ the molecules comprising the medium become polarized, giving rise to a dipole moment per unit volume described by the polarization vector $\mathbf{P}$ in units of C/m². It can be assumed that the polarization is linearly related to the field strength by a relation of the form

$$\mathbf{P} = \varepsilon_0 \chi_e \mathbf{E}, \tag{1.4}$$

where χ_e is a dimensionless quantity called the dielectric susceptibility. The relationship between the vectors $\mathbf{D}$ and $\mathbf{E}$ can then be given in terms of the polarization by the expression

$$\mathbf{D} = \varepsilon_0 \mathbf{E} + \mathbf{P}, \tag{1.5}$$

where consistency between Eqs. (1.3a) and (1.5) requires that

$$\kappa_e = \frac{\varepsilon}{\varepsilon_0} = 1 + \chi_e. \tag{1.6}$$

The quantity μ appearing in Eq. (1.3b) is called the absolute magnetic permeability and has the units H/m, where the henry is the unit of inductance and is equivalently expressed as H = Wb/A. For fields in free space, the permeability has the value $\mu_0 = 4\pi \times 10^{-7}$ H/m, and thus a relative permeability $\kappa_m = \mu/\mu_0$ can be defined for arbitrary media. A magnetic material responds to an applied magnetic field $\mathbf{H}$ by developing a magnetic dipole moment, which is attributable to the internal currents (called Amperian currents) induced by the external field. Thus a magnetization vector $\mathbf{M}$ can be defined as the induced magnetic dipole moment per unit volume in units of A/m. For most materials (the notable exception being the ferromagnetics) the magnetization can be assumed to be linearly related to the magnetic field by the relation

$$\mathbf{M} = \chi_m \mathbf{H}, \tag{1.7}$$

where χ_m is the magnetic susceptibility. The relationship between the vectors **B** and **H** can be given in terms of the magnetization by the expression

$$\mathbf{B} = \mu_0 (\mathbf{H} + \mathbf{M}). \tag{1.8}$$

The magnetic induction **B** is thus composed of a contribution $\mu_0 \mathbf{M}$ resulting from Amperian currents and a contribution $\mu_0 \mathbf{H}$ which, by Eq. (1.1b), may have an external current as its source. Consistency between relations (1.3b) and (1.8) requires that

$$\kappa_m \equiv \frac{\mu}{\mu_0} = 1 + \chi_m. \tag{1.9}$$

Materials for which $\chi_m < 0$ are called diamagnetic, and materials for which $\chi_m > 0$ are called paramagnetic. In a paramagnetic material, the Amperian currents act so as to aid the external currents, whereas in a diamagnetic material the Amperian currents act so as to oppose the external currents.

Finally, the quantity σ appearing in the constitutive relation (1.3c) (which can be considered to be a form of Ohm's law) is called the electric conductivity and is measured in units of $[\Omega^{-1} \text{ m}^{-1}]$, where the ohm is the unit of electrical resistance and is equivalently expressed as $\Omega = \text{V/A}$.

1.2 ENERGY AND MOMENTUM RELATIONS

It is possible to make a consistent assignment of energy to the fields that appear in Maxwell's equations. To establish the form of this relation, we first note the following vector identity:

$$\nabla \cdot (\mathbf{E} \times \mathbf{H}) = \mathbf{H} \cdot (\nabla \times \mathbf{E}) - \mathbf{E} \cdot (\nabla \times \mathbf{H}).$$

The right-hand side of this expression can be rewritten using the Maxwell equations (1.1a) and (1.1b) and the constitutive equations (1.3a) and (1.3b) to give the result

$$\nabla \cdot (\mathbf{E} \times \mathbf{H}) + \frac{\partial}{\partial t} \frac{1}{2} (\mathbf{D} \cdot \mathbf{E} + \mathbf{B} \cdot \mathbf{H}) = -\mathbf{J} \cdot \mathbf{E}. \tag{1.10}$$

This equation can be regarded as a statement of energy balance in which the quantity $\mathbf{J} \cdot \mathbf{E}$ gives the rate per unit volume at which energy is lost by the field due to Joule heating. The quantity

$$\mathbf{S} \equiv \mathbf{E} \times \mathbf{H} \tag{1.11}$$

is known as the Poynting vector and gives the rate at which electromagnetic energy passes through a unit area whose normal is in the direction of **S**. The quantity

$$u = \tfrac{1}{2}(\mathbf{D} \cdot \mathbf{E} + \mathbf{B} \cdot \mathbf{H}) \tag{1.12}$$

can be regarded as the energy per unit volume stored in the field. It is likewise possible to assign a momentum density to the electromagnetic field, which can be expressed as*

$$\mathbf{g} = \mu\varepsilon(\mathbf{E} \times \mathbf{H}) = \mathbf{D} \times \mathbf{B}. \tag{1.13}$$

1.3 THE WAVE EQUATION AND INFINITE PLANE WAVES

Maxwell's equations predict the existence of propagating electromagnetic waves, of which light is one example. We shall consider the properties of such radiation for the case of a uniform medium that is free of sources (i.e., $\rho = 0$, and the only currents are those given by Ohm's law, $\mathbf{J} = \sigma\mathbf{E}$) and for which constitutive relations of the form (1.3) may be assumed. By taking the curl of the Maxwell equation (1.1a), one obtains the equation

$$\nabla \times \nabla \times \mathbf{E} = -\mu\frac{\partial}{\partial t}(\nabla \times \mathbf{H}).$$

The left-hand side of this equation can be identically expressed as $\nabla(\nabla \cdot \mathbf{E}) - \nabla^2\mathbf{E}$, where the first term vanishes, [according to Eqs. (1.1d) and (1.3a)], since it has been assumed that $\rho = 0$. If Eq. (1.1b) is introduced into the right-hand side, one obtains the wave equation

$$\nabla^2\mathbf{E} = \mu\sigma\frac{\partial \mathbf{E}}{\partial t} + \varepsilon\mu\frac{\partial^2 \mathbf{E}}{\partial t^2}. \tag{1.14a}$$

An analogous calculation gives the wave equation for the magnetic intensity in the same form:

$$\nabla^2\mathbf{H} = \mu\sigma\frac{\partial \mathbf{H}}{\partial t} + \varepsilon\mu\frac{\partial^2 \mathbf{H}}{\partial t^2}. \tag{1.14b}$$

---

*For the case of an electromagnetic wave in a material medium, this equation gives the total momentum density, including both the momentum of the electromagnetic field and the mechanical momentum associated with the response of the medium. It is generally accepted that the momentum density associated with the field only is given by $g = \mathbf{E} \times \mathbf{H}$. This point is discussed in Section 6.9 of *Classical Electrodynamics* by J. D. Jackson, Wiley, New York, 1975 and in Section 69 of *Electrodynamics of Continuous Media* by L. D. Landau and E. M. Lifshitz, Pergamon, New York, 1960.

Dielectric Media

We shall first restrict our attention to the case of zero conductivity $(\sigma = 0)$ and determine the form of the solutions to the wave equation for this case. We seek solutions in the form of plane waves

$$\mathbf{E}(\mathbf{r}, t) = E_0 \hat{\mathbf{e}}\, e^{i(\mathbf{k} \cdot \mathbf{r} - \omega t)}, \tag{1.15a}$$

$$\mathbf{H}(\mathbf{r}, t) = H_0 \hat{\mathbf{h}}\, e^{i(\mathbf{k} \cdot \mathbf{r} - \omega t)}, \tag{1.15b}$$

where E_0 and H_0 are complex field amplitudes and $\hat{\mathbf{e}}$ and $\hat{\mathbf{h}}$ are complex polarization unit vectors, and where it is implicitly assumed that the physical fields correspond to the real parts of these expressions. It is appropriate to restrict our attention to plane-wave solutions because at large distances from a source the emitted wave front will be nearly planar and because, as a consequence of Fourier's theorem, an arbitrary radiation field can be decomposed into a linear superposition of plane-wave components. If the expressions (1.15) are to be solutions of the wave equations (1.14) for $\sigma = 0$, the quantities $\mathbf{k}$ and ω must be related by

$$k \equiv |\mathbf{k}| = \omega (\varepsilon \mu)^{1/2}. \tag{1.16}$$

This dispersion relation implies that the phase velocity of the radiation is given by $(\varepsilon \mu)^{-1/2}$, which for propagation through free space is conventionally denoted by

$$c = (\varepsilon_0 \mu_0)^{-1/2} \simeq 2.998 \times 10^8 \text{ m/s}. \tag{1.17}$$

The refractive index N can be defined by

$$N = \left(\frac{\varepsilon \mu}{\varepsilon_0 \mu_0} \right)^{1/2} = (\kappa_e \kappa_m)^{1/2}. \tag{1.18}$$

With this definition, the dispersion relation (1.16) can be expressed as

$$k = N \frac{\omega}{c}, \tag{1.19}$$

leading to an expression for the wavelength of the form

$$\lambda = \frac{2\pi c}{N\omega}. \tag{1.20}$$

Additional constraints are placed on the form of the plane-wave solutions by requiring that they satisfy Maxwell's equations. In particular, the requirements that $\nabla \cdot \mathbf{E} = 0$ and $\nabla \cdot \mathbf{H} = 0$ demand that

$$\mathbf{k} \cdot \hat{\mathbf{e}} = 0, \qquad \mathbf{k} \cdot \hat{\mathbf{h}} = 0, \qquad (1.21)$$

implying that both $\hat{\mathbf{e}}$ and $\hat{\mathbf{h}}$ are perpendicular to $\mathbf{k}$. In addition, if the solutions (1.15) are to satisfy the curl $\mathbf{E}$ equation (1.1a), it must be required that

$$(\mathbf{k} \times \hat{\mathbf{e}}) E_0 = \omega\mu\hat{\mathbf{h}}H_0. \qquad (1.22)$$

This implies that the vectors $\mathbf{k}, \hat{\mathbf{e}}$, and $\hat{\mathbf{h}}$ are in fact mutually perpendicular, as shown in Figure 1.1. Furthermore, the quantities ω and $|\mathbf{k}|$ can be eliminated from this expression, through use of Eq. (1.16), to show that the field amplitudes are related by the expression

$$\frac{E_0}{H_0} = \sqrt{\frac{\mu}{\varepsilon}} \equiv Z. \qquad (1.23)$$

Here Z is called the wave impedance and is clearly a real number, indicating that the fields E and H oscillate in phase. For waves propagating in vacuum, this quantity takes the particular value $Z_0 = \sqrt{\mu_0/\varepsilon_0} \simeq 377$ Ω, called the impedance of free space.

The Poynting vector $\mathbf{S} = \mathbf{E} \times \mathbf{H}$ for such an electromagnetic wave will exhibit rapid oscillations at the angular frequency 2ω. In many circum-

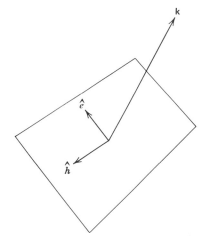

Figure 1.1. Propagation vector $\mathbf{k}$ and the polarization vectors $\hat{\mathbf{e}}$ and $\hat{\mathbf{h}}$.

stances, however, one is interested only in the time average of the Poynting vector, which gives the time-averaged rate at which energy is transported by the fields. The calculation of the time average of quantities such as S which depend on the real parts of complex quantities is greatly simplified by the following general rule: If A and B show the same harmonic time dependence

$$A = A_0 e^{-i\omega t}, \qquad B = B_0 e^{-i\omega t},$$

then the time average of the product of the real parts of these quantities is given by

$$\langle (\operatorname{Re} A)(\operatorname{Re} B) \rangle_t = \tfrac{1}{2} \operatorname{Re}(A_0 B_0^*). \tag{1.24}$$

Thus the time-averaged Poynting vector is given by

$$\langle S \rangle_t = \tfrac{1}{2} E_0 H_0^* \hat{k}, \tag{1.25a}$$

$$= \frac{1}{2} \sqrt{\frac{\varepsilon}{\mu}} |E_0|^2 \hat{k}, \tag{1.25b}$$

where $\hat{k} = k/|k|$ is a unit vector in the propagation direction and where Eq. (1.23) has been used to obtain the second line. A time average of expression (1.12) for the energy density can likewise be performed, leading to the result

$$\langle u \rangle_t = \tfrac{1}{4} \left(\varepsilon |E_0|^2 + \mu |H_0|^2 \right).$$

Through use of Eq. (1.23), it is seen that both terms in this expression are equal in magnitude, showing that equal energies reside in the electric and magnetic fields and allowing the expression to be simplified to

$$\langle u \rangle_t = \tfrac{1}{2} \varepsilon |E_0|^2. \tag{1.26}$$

Conducting Media

For the case in which the conductivity σ in Eqs. (1.14) is not identically zero, plane-wave solutions of the form (1.15) still exist if the propagation vector k is allowed to become complex. The requirement that Eqs. (1.15a) and (1.15b) satisfy Eqs. (1.14a) and (1.14b), respectively, leads to a dispersion relation of the form

$$k = N \frac{\omega}{c}, \tag{1.27}$$

where $N = N' + iN''$ is now the complex refractive index given by the expression

$$N^2 = \frac{\mu\varepsilon}{\mu_0\varepsilon_0}\left(1 + \frac{i\sigma}{\varepsilon\omega}\right). \tag{1.28}$$

The real and imaginary parts of N are given explicitly by the relations

$$\binom{N'}{N''} = \left(\frac{\mu\varepsilon}{\mu_0\varepsilon_0}\right)^{1/2}\left[\frac{1}{2}\sqrt{1 + \left(\frac{\sigma}{\varepsilon\omega}\right)^2} \pm \frac{1}{2}\right]^{1/2}. \tag{1.29}$$

The plane-wave solutions (1.15) can be written in terms of the complex refractive index and take the form

$$\mathbf{E}(\mathbf{r}, t) = E_0\hat{\mathbf{e}}\, e^{-N''(\omega/c)(\hat{\mathbf{k}}\cdot\mathbf{r})}e^{i[N'(\omega/c)\hat{\mathbf{k}}\cdot\mathbf{r} - \omega t]}, \tag{1.30a}$$

$$\mathbf{H}(\mathbf{r}, t) = H_0\hat{\mathbf{h}}\, e^{-N''(\omega/c)(\hat{\mathbf{k}}\cdot\mathbf{r})}e^{i[N'(\omega/c)\hat{\mathbf{k}}\cdot\mathbf{r} - \omega t]}, \tag{1.30b}$$

where, as before, $\hat{\mathbf{k}}$ represents a unit vector in the direction of the propagation vector $\mathbf{k}$. We note that the field strengths decrease exponentially with the distance $\hat{\mathbf{k}}\cdot\mathbf{r}$ through which the wave has propagated. This spatial variation can be described by introducing the absorption coefficient

$$\alpha = 2N''\frac{\omega}{c} = \frac{4\pi N''}{\lambda}, \tag{1.31}$$

which is defined as the inverse of the propagation distance through which the Poynting vector falls to $1/e$ of its initial value, so that

$$\mathbf{S}(\mathbf{r}) = \mathbf{S}(0)e^{-\alpha(\hat{\mathbf{k}}\cdot\mathbf{r})} \tag{1.32}$$

The inverse of the absorption coefficient is known as the penetration depth.

As in the case of dielectric media, the vectors $\mathbf{k}$, $\hat{\mathbf{e}}$, and $\hat{\mathbf{h}}$, must be mutually orthogonal, but now the condition (1.22), imposed by requiring that $\mathbf{E}(\mathbf{r}, t)$ and $\mathbf{H}(\mathbf{r}, t)$ satisfy the Maxwell equation (1.1a), leads to fields related by an expression of the form

$$\frac{E_0}{H_0} = \sqrt{\frac{\mu}{\varepsilon}}\,\frac{1}{\left(1 + \dfrac{i\sigma}{\varepsilon\omega}\right)^{1/2}}. \tag{1.33}$$

A consequence of this result is that in the limit of highly conducting media $(\sigma/\varepsilon\omega \gg 1)$ the magnetic intensity H lags the electric field E by $45°$, and most of the energy of the wave resides in the magnetic field.

The treatment presented here is based on the assumption that the quantities ε and σ are real, and that absorption results from a nonzero value of the conductivity σ. In certain circumstances, however, such as those involving absorption by bound charges, it is more natural to treat absorption as resulting from the imaginary part of a complex susceptibility χ_e or permittivity ε, with the conductivity taken as zero. In a sense, these two methods present alternative viewpoints for describing identical physical processes. The treatment presented here correctly describes absorption phenomena if the quantity ε is treated as a complex number in Eqs. (1.27) and (1.28) and Eqs. (1.30) through (1.33).

1.4 BOUNDARY CONDITIONS

Many problems in optics require the solution of Maxwell's equations under conditions whereby some physical property, such as the dielectric constant κ_e or conductivity σ, changes discontinuously between neighboring regions. In such cases, it is necessary to determine the boundary conditions that the fields must satisfy at the discontinuity. The form of these boundary conditions is found from Maxwell's equations (1.1) by a limiting procedure described in standard books on electromagnetic theory (see, e.g., those listed in the bibliography at the end of this chapter.) Defining $\hat{n}_{21}$ to be a unit vector pointing from region 1 into region 2, and labeling the fields in these regions by subscripts 1 and 2, the boundary conditions are given by

$$\hat{n}_{21} \times (\mathbf{E}_2 - \mathbf{E}_1) = 0, \tag{1.34a}$$

$$\hat{n}_{21} \times (\mathbf{H}_2 - \mathbf{H}_1) = \mathbf{J}^{(s)}, \tag{1.34b}$$

$$\hat{n}_{21} \cdot (\mathbf{D}_2 - \mathbf{D}_1) = \rho^{(s)}, \tag{1.34c}$$

$$\hat{n}_{21} \cdot (\mathbf{B}_2 - \mathbf{B}_1) = 0. \tag{1.34d}$$

In these equations, $\rho^{(s)}$ denotes the surface charge density, that is, the charge per unit area stored on the boundary between the two surfaces. The units of $\rho^{(s)}$ are C/m^2. The quantity $\mathbf{J}^{(s)}$ denotes the surface current density; it measures the flow of current along the surface in units of A/m. For the case of a discontinuity between two dielectric layers, both $\mathbf{J}^{(s)}$ and $\rho^{(s)}$ must vanish. In this case, the boundary conditions can be interpreted as requiring that the normal components of **D** and **B** and the tangential components of **E** and of **H** be continuous at the boundary. Another common special case is that of a discontinuity between a dielectric and a perfect conductor. The field **E** must, by Eq. (1.3c), vanish inside the perfect conductor to avoid the existence of an infinite current density **J**, and therefore, $\mathbf{D} = \varepsilon\mathbf{E}$ must also vanish. Thus, by Eq. (1.1a), no time-varying **B** field can exist inside a perfect metal, and by Eq. (1.3b) no **H** field can exist either. Therefore, for the case of time-varying fields, only the normal components of **D** and **E** and the tangential components of **B** and **H** are permitted at the surface of the conductor.

BIBLIOGRAPHY

M. Born and E. Wolf, *Principles of Optics*, Pergamon, New York, 1975, Chapters 1 and 2.

R. P. Feynman, R. B. Leighton, and M. Sands, *Lectures on Physics*, Addison-Wesley, Reading, Mass., 1964, especially Volume II.

J. D. Jackson, *Classical Electrodynamics*, Wiley, New York, 1975.

L. D. Landau and E. M. Lifshitz, *Classical Theory of Fields*, Addison-Wesley, Reading, Mass., 1962.

L. D. Landau and E. M. Lifshitz, *Electrodynamics of Continuous Media*, Addison-Wesley, Reading, Mass., 1960.

W. K. H. Panofsky and M. Phillips, *Classical Electricity and Magnetism*, Addison-Wesley, Reading, Mass., 1955.

E. M. Purcell, *Electricity and Magnetism*, McGraw-Hill, New York, 1965.

J. C. Slater and N. H. Frank, *Electromagnetism*, Dover, New York, 1969.

A. Sommerfeld, *Electrodynamics*, Academic, New York, 1964.

PROBLEMS

1 Calculate the penetration depth into copper at room temperature of electromagnetic radiation at frequencies of 60 Hz, 1 MHz, and 500 THz.

2 Consider a coaxial cable whose inner conductor has radius a and whose outer conductor has an inner radius b, the space between the conductors being filled with a lossless dielectric. Assume the conductors have negligible resistivity. A voltage signal of the form

$$V(z, t) = V_0 e^{i(kz - \omega t)}$$

propagates along the cable.

By solving Maxwell's equations with appropriate boundary conditions, find the field strengths $\mathbf{E}$ and $\mathbf{H}$ within the dielectric and the current I carried by the cable, assuming that both $\mathbf{E}$ and $\mathbf{H}$ are transverse (i.e., that a TEM wave is excited). Determine the characteristic impedance $Z = V/I$ of the cable, the wave impedance $Z' = E/H$ of the cable, the power carried by the cable, and the functional dependence of k on ω.

3   A commonly encountered type of coaxial cable known as RG58 A/U has a characteristic impedance of 50 Ω. The inner conductor has a diameter of 0.0375 in., and is surrounded by an insulating layer whose outer diameter is 0.120 in. The outer conductor is a tinned copper braid whose outer diameter is 0.150 in., and the entire cable is enclosed in a polyvinyl-chloride jacket of outer diameter 0.199 in. What is the dielectric constant of the insulating layer?

4   A coaxial cable of characteristic impedance Z_1 is spliced to another cable of impedance Z_2. What is the power reflection coefficient for an electromagnetic wave incident on the interface?

(This problem illustrates the importance of properly terminating transmission lines used in detection systems designed to operate at high speeds. A cable is said to be *terminated* by a resistance R connecting its two conductors. If R is equal to the characteristic impedance Z of the cable, all of the power incident on R will be absorbed in R and none will be reflected. However, if $R \neq Z$, some of the power will be reflected back toward the source. A further reflection at the source can lead to "ringing" in the cable, which severely limits the temporal response of the system.)

5   A dielectric waveguide has the form of a cylinder of radius R. The refractive index of the cylinder is n, while that of the surrounding medium is $n_0 < n$. What is the field distribution of the propagating modes within the waveguide, and what is the relation between ω and k for each such mode?

6   Find $\mathbf{E}(\mathbf{r}, t)$ and $\mathbf{H}(\mathbf{r}, t)$ for the normal modes of a rectangular cavity with perfectly conducting walls.

2

Radiometry

Radiometry deals with the problem of measuring the energy content of optical radiation fields and of determining how this energy flows through optical systems. The present chapter contains an elementary discussion of the subject of radiometry; a discussion of the more advanced topics of radiometry will be postponed until Chapter 5 so that the subject matter of Chapters 3 and 4, dealing with blackbody radiation, can be included in the treatment.

The proper formalism for discussing energy transport by radiation fields depends critically on the coherence of the radiation. Optical radiation can be classified as fully coherent, partially coherent, or incoherent. Chapter 1 presented a discussion of electromagnetic plane waves. Plane waves are an example of fully coherent radiation, since the amplitude and phase of such waves are given by Eq. (1.15) and thus are fixed for all of space. The radiometric properties of fully coherent radiation are specified by a knowledge of the electromagnetic fields through use of Eqs. (1.11) and (1.12) for the Poynting vector and the electromagnetic energy density.

The traditional problem of radiometry, however, is to specify the energy flow of essentially incoherent radiation fields, and the ensuing discussion in this and the following chapters deals with the radiometry of essentially incoherent sources. Thermal sources, for example, produce essentially incoherent radiation. For such radiation, the amplitude and phase of the optical fields fluctuate randomly in space. In the case of a thermal source, this randomness occurs because the various elements of the source radiate independently of one another. The optical field near the source is thus the result of the superposition of a great number of independent waves, which combine in a stochastic manner.

The laws of traditional radiometry are based on several assumptions. The first assumption, as discussed earlier, is that the sources are incoherent. In addition, it is assumed that the propagation of this radiation through empty space and through optical systems can be treated using the laws of geometrical optics. In particular, it is assumed that radiant energy is transported along the direction of the rays of geometrical optics. Thus the laws of radiometry cannot treat those problems where interference effects or diffraction effects are dominant, since these effects can result in the flow of energy in directions other than those of geometrical rays. Finally, it is assumed that the energy of the optical fields is conserved as long as the field propagates through transparent (i.e., nonabsorbing) media.

2.1 RADIOMETRIC QUANTITIES

It has proven useful to define a number of radiometric quantities to describe the energy content of incoherent radiation fields. These quantities are listed in Table 2.1 and are described in detail in this section. In the past there has been considerable disagreement regarding the nomenclature used for the radiometric units. The names used in this book are those most often used by present-day workers. Considerable care should be exercised, however, when comparing the results of various authors, since the same word sometimes is used to denote different physical properties.

The total energy contained in a radiation field or the total energy delivered to a given receiver by such a radiation field is called the radiant energy Q. The energy density is defined as

$$u = \frac{dQ}{dV}, \tag{2.1}$$

where dQ is the radiant energy contained in a volume element dV of the radiation field. The radiant flux Φ, equivalently referred to as the power P, is the rate dQ/dt at which radiant energy is transferred from one region to another by the radiation field.

Table 2.1. **Radiometric Quantities and Units**

Quantity	Symbol	Definition	Unit
Radiant energy	Q	$\int \Phi \, dt$	J
Radiant energy density	u	dQ/dV	J/m^3
Radiant flux (power)	Φ, P	dQ/dt	W
Radiant exitance	M	$d\Phi/dA$	W/m^2
Irradiance	E	$d\Phi/dA$	W/m^2
Radiance	L	$d^2\Phi/dA_{\text{proj}} \, d\Omega$	W/m^2 sr
Radiant intensity	I	$d\Phi/d\Omega$	W/sr

The radiant exitance M is the flux per unit area leaving the surface of a source of radiation and is equivalently expressed as

$$M = \frac{d\Phi}{dA},$$
(2.2)

where $d\Phi$ is the flux leaving a source element of area dA. Similarly, the irradiance E, also called the radiant incidence is the flux per unit area received by a real or imaginary surface element and is expressed as

$$E = \frac{d\Phi}{dA}.$$
(2.3)

Since the remaining radiometric quantities involve the concept of flux per unit solid angle, it is worthwhile to recall first the definition of solid angle. By definition, a cone is the volume swept out when a straight line passing through a fixed vertex is moved through every point on a closed, nonintersecting curve, as illustrated in Fig. 2.1. The solid angle Ω in units of steradians (sr) subtended by a cone is given by the expression

$$\Omega = \frac{a}{r^2},$$
(2.4)

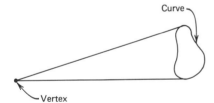

Figure 2.1. Definition of a cone.

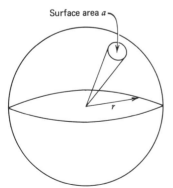

Figure 2.2. Definition of solid angle.

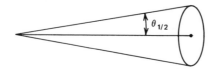

Figure 2.3. Right circular cone.

where a is the area intercepted by the cone on the surface of a sphere of radius r centered on the cone vertex, as shown in Fig. 2.2. The solid angle subtended by an entire sphere is thus equal to 4π sr, and the solid angle subtended by a right circular cone of half vertex angle $\theta_{1/2}$, shown in Fig. 2.3, is given by the expression

$$\Omega = 4\pi \sin^2 \tfrac{1}{2}\theta_{1/2}. \tag{2.5}$$

The radiance L is the flux per unit projected area per unit solid angle leaving a source or, in general, any reference surface. If $d^2\Phi$ is the flux emitted into a solid angle $d\Omega$ by a source element of projected area dA_{proj}, the radiance is defined as

$$L = \frac{d^2\Phi}{dA_{\text{proj}}\, d\Omega}, \tag{2.6a}$$

where, as shown in Fig. 2.4, the projected area is given by

$$dA_{\text{proj}} = dA \cos\theta, \tag{2.6b}$$

where θ is the angle between the outward surface normal of the area element dA and the direction of observation. We shall see later that the radiance

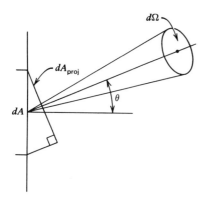

Figure 2.4. Geometrical quantities appearing in the definition of the radiance.

plays a special role in radiometry because it is conserved for propagation through a lossless optical system. The radiance is sometimes referred to as the brightness or specific intensity.

The radiant intensity I is the flux per unit solid angle emitted by an entire source in a given direction; therefore, it can be expressed as

$$I = \frac{d\Phi}{d\Omega},\qquad(2.7)$$

where $d\Phi$ is the flux emitted into the solid angle $d\Omega$. Although a radiant intensity can be defined for any source, this quantity is a particularly useful description of the radiation produced by a point source, that is, a source whose linear dimensions are much less than the distance from the source to the region of observation. Figure 2.5 illustrates the illumination of a surface element by a point source of intensity I. This surface element subtends a solid angle given according to Eq. (2.4) by $d\Omega = dA \cos\theta/r^2$. By Eq. (2.7), the flux hitting the surface element is given by $d\Phi = I\,dA\cos\theta/r^2$, and, according to Eq. (2.3), the irradiance at dA is given by

$$E = \frac{d\Phi}{dA} = \frac{I\cos\theta}{r^2}.\qquad(2.8)$$

This result illustrates one form of Lambert's cosine law in that the irradiance decreases with the angle of incidence as $\cos\theta$. Equation (2.8) also illustrates the inverse-square law in that the irradiance decreases with distance as $1/r^2$. We shall see later in this chapter that, in general, the inverse-square law is valid only for point sources.

It should be noted that the word *intensity* has a very different meaning in the field of physical optics, where it refers to the magnitude of the Poynting vector and thus corresponds more closely to *irradiance* in the nomenclature of radiometry.

2.2 SPECTRAL RADIOMETRIC QUANTITIES

The radiometric quantities defined in Section 2.1 refer to the total energy of the radiation field with no regard to the spectral composition of the

Figure 2.5.   Illumination of a surface element by a point source.

radiation. In many situations it is necessary to introduce radiometric quantities that explicitly take into account this spectral composition. Thus we can define

$$\Phi_\lambda \, d\lambda = \text{flux in the wavelength interval } \lambda \text{ to } \lambda + d\lambda$$

so that

$$\Phi = \int_0^\infty \Phi_\lambda \, d\lambda. \tag{2.9}$$

For those circumstances in which a frequency distribution is preferred, we can define

$$\Phi_\nu \, d\nu = \text{flux in the frequency interval } \nu \text{ to } \nu + d\nu \tag{2.10}$$

where

$$\Phi = \int_0^\infty \Phi_\nu \, d\nu.$$

The quantities Φ_ν and Φ_λ can be related by noting that if $d\nu$ is the frequency interval corresponding to the wavelength interval $d\lambda$, it must be true that

$$\Phi_\nu \, d\nu = \Phi_\lambda \, d\lambda,$$

since both sides of this equation measure the same physical quantity. Since $\nu = c/\lambda$, we obtain

$$\Phi_\lambda = \Phi_\nu \left| \frac{d\nu}{d\lambda} \right| = \Phi_\nu \frac{c}{\lambda^2}, \tag{2.11}$$

which can be expressed in the more symmetric form

$$\lambda \Phi_\lambda = \nu \Phi_\nu. \tag{2.12}$$

Spectral versions of the other radiometric quantities can be defined similarly.

2.3   RADIANCE CONSERVATION FOR FREE PROPAGATION

An important law of radiometry known as the radiance theorem states that radiance is conserved for propagation through a lossless optical system.

At this point, we prove this result for the important special case of propagation through a homogeneous medium. The radiance theorem will be proved in general in Chapter 5.

Figure 2.6 shows two area elements dA_0 and dA_1 separated by the distance r. It is useful to define two solid angles $d\Omega_0$ and $d\Omega_1$ as follows:

$$d\Omega_0 = \text{solid angle subtended by } dA_1 \text{ at } dA_0$$

$$= \frac{dA_1 \cos\theta_1}{r^2}; \tag{2.13}$$

$$d\Omega_1 = \text{solid angle subtended by } dA_0 \text{ at } dA_1$$

$$= \frac{dA_0 \cos\theta_0}{r^2}. \tag{2.14}$$

If L_0 is the radiance of the radiation field measured at dA_0 in the direction of dA_1, the flux transferred from dA_0 to dA_1 is given, as a consequence of the definition of radiance (2.6), by the expression

$$d^2\Phi = L_0(dA_0 \cos\theta_0)(d\Omega_0). \tag{2.15}$$

The radiance L_1 of the radiation field measured at dA_1 in this same direction is given by

$$L_1 = \frac{d^2\Phi}{dA_1 \cos\theta_1 \, d\Omega_1}. \tag{2.16}$$

where $d^2\Phi$ is given by Eq. (2.15) and where the solid angle element is given by Eq. (2.14), since the flux must leave the surface dA_1 in a solid angle equal to that from which it arrived. Equation (2.16) can be simplified through use of Eqs. (2.13)–(2.15) to give the result

$$L_1 = L_0, \tag{2.17}$$

showing that radiance is conserved for freely propagating radiation.

Figure 2.6. Geometrical construction used to demonstrate conservation of radiance.

An additional interesting result can be obtained by using Eqs. (2.13) and (2.14) to recast the expression (2.15) for the transmitted flux into the form

$$d^2\Phi = L_0 \cos\theta_0 \cos\theta_1 \frac{dA_0\, dA_1}{r^2}$$

$$= L_0(dA_1 \cos\theta_1)\, d\Omega_1. \tag{2.18}$$

Comparison of Eqs. (2.15) and (2.18) shows that the transmitted flux can be obtained by multiplying the source radiance by a projected-area–solid-angle product measured at either the source or the receiver surface. This result is useful, because it allows one to adopt the point of view of either the source or the receiver in performing radiometric calculations.

2.4 LAMBERTIAN SOURCES

It is an empirical fact that most incoherent radiation sources (i.e., emitters or scatterers) produce radiation whose radiance is approximately independent of the angle of observation. A Lambertian source is by definition one whose radiance is completely independent of viewing angle.

Figure 2.7 shows a planar Lambertian source of area A_0 and uniform radiance L_0. By comparing the definitions (2.6) and (2.7) of the radiance and the intensity, one can obtain the following expression for the intensity of the source in the direction θ:

$$I = \int_{\text{source}} L_0 \cos\theta\, dA.$$

Since the integrand is independent of position, this integral is evaluated trivially to give

$$I = I_0 \cos\theta, \tag{2.19}$$

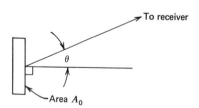

Figure 2.7. Lambertian source of area A_0.

where I_0 has been used to designate the product $L_0 A_0$. Equation (2.19) constitutes another form of Lambert's cosine law and illustrates the decrease of source intensity with observation angle for a planar source of finite size. It should be noted that this decrease is due entirely to the decreased projected area of the source.

An additional property of Lambertian sources is illustrated by the experimental setup shown in Fig. 2.8. An opening of area dA_0 in screen S is irradiated by a Lambertian source which can be rotated so as to vary the angle θ between the source normal and the screen normal. The flux hitting an area element dA_1 that is parallel to screen S and centered on the opening is measured as a function of the angle θ. From the definition (2.6) of radiance, the flux hitting the element dA_1 is given by

$$d^2\Phi = L_0 \, dA_{\text{proj}} \, d\Omega. \tag{2.20}$$

Here the projected area element is given by $dA_{\text{proj}} = dA_0$ for any angle θ, and the solid angle is given by $d\Omega = dA_1/r^2$. Thus the expression (2.20) becomes

$$d^2\Phi = \frac{L_0 \, dA_0 \, dA_1}{r^2}, \tag{2.21}$$

which is independent of the orientation angle θ.

Except at grazing incidence, white paper is a good approximation to a Lambertian scatterer. One may verify that the subjective brightness of a piece of paper does not change as the paper is viewed at a constant distance from a variety of angles, although of course the solid angle subtended by the paper changes. These observations are consistent with Eq. (2.20), which

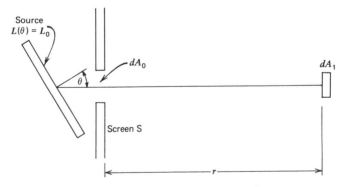

Figure 2.8. Measurement involving a Lambertian source.

shows that the same flux is incident on the eye from each projected area element, and with Eq. (2.19), which shows that the total flux hitting the eye decreases as the projected area of the source decreases.

For Lambertian sources, a particularly simple relation holds between the radiance and the radiant exitance. From the definitions (2.2) and (2.6) of these two quantities, it follows that

$$M = \int L \cos \theta \, d\Omega$$

$$= 2\pi L \int_0^{\pi/2} \sin \theta \cos \theta \, d\theta$$

$$= 2\pi L \int_0^1 \cos \theta \, d(\cos \theta)$$

$$= \pi L. \tag{2.22}$$

It will be shown in the next chapter that a small opening in a blackbody cavity constitutes a theoretically perfect Lambertian source.

Figure 2.9 shows a source that is extremely non-Lambertian: an illuminated piece of white paper baffled by black slats. When viewed at normal incidence to the paper, the white paper is visible and the object has a large radiance. However, when viewed at oblique incidence, only the black slats are visible, and the object has a small radiance. Another example of a non-Lambertian source is a grass lawn. If the grass is viewed from directly overhead, some of the earth under the grass will be visible. However, if the lawn is viewed at an oblique angle, only the green blades of grass will be visible. Thus the visual appearance of a grass lawn depends on the angle at which it is viewed. It has been suggested (perhaps facetiously) that this phenomenon is the origin of the expression that the grass is always greener on the other side of the fence.

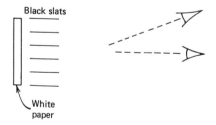

Figure 2.9. Non-Lambertian source.

2.5 FINITE SOURCES

Many of the fundamental radiometric quantities are defined in terms of infinitesimal area elements. Thus, in order to treat problems dealing with finite sources, it is often necessary to perform integrations of these quantities over the surface of the source. There are a number of cases of practical importance for which these integrations can be performed in closed form, and we discuss two examples in the following sections.

Disk Lambertian Source

We wish to calculate the irradiance produced by a disk Lambertian source of radius R and uniform radiance L at an area element dA_1 which lies parallel to the surface of the disk and is located axially at a distance z from the center of the disk, as shown in Fig. 2.10. An annular element of area on the source is given by

$$dA_0 = 2\pi z^2 \frac{\sin\theta\, d\theta}{\cos^3\theta}, \qquad (2.23)$$

and the element of solid angle subtended by dA_1 from any point on dA_0 is given by

$$d\Omega_0 = \frac{dA_1 \cos\theta}{(z/\cos\theta)^2}. \qquad (2.24)$$

Thus the flux transferred from dA_0 to dA_1 is given, in accordance with the definition of the radiance (2.6), by the expression

$$d^2\Phi = 2\pi L\, dA_1 \sin\theta \cos\theta\, d\theta, \qquad (2.25)$$

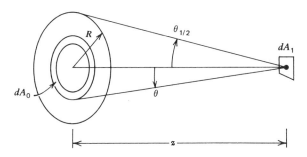

Figure 2.10. Disk Lambertian source irradiates the area element dA_1.

and the irradiance at dA_1 is given by

$$E = \frac{d\Phi}{dA_1} = \int_0^{\theta_{1/2}} 2\pi L \sin\theta \cos\theta \, d\theta$$

$$= \pi L \sin^2\theta_{1/2} = \pi L \left(\frac{R^2}{R^2 + z^2} \right), \tag{2.26}$$

where

$$\theta_{1/2} = \arctan\frac{R}{z} \tag{2.27}$$

is the half angle subtended by the disk at the point of observation.

We note that in the limit $z \ll R$ the irradiance E approaches the value πL, which by Eq. (2.22) is the radiant exitance of a Lambertian source. We also note that in the opposite limit $z \gg R$ the irradiance E approaches the value $\pi R^2 L/z^2$, and for this case the irradiance obeys the inverse-square law. In this limit the disk can be considered to be a point source of intensity $I = \pi R^2 L$, and the irradiance can then be expressed by the general result $E = I/r^2$ derived for point sources as Eq. (2.8).

Spherical Lambertian Source

In this case, we wish to determine the irradiance E on an area element positioned a distance r from the center of a spherical Lambertian source of uniform radiance L, as shown in Fig. 2.11. Although it is possible to determine this irradiance by performing an integration over the surface of the source, the calculation is greatly simplified if explicit use is made of the spherical symmetry of the emitted radiation. The radiant exitance of a Lambertian source, is given by Eq. (2.22) as

$$M = \pi L.$$

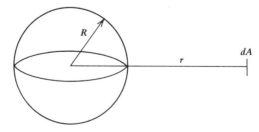

Figure 2.11.   Spherical Lambertian source irradiates the area element dA.

Thus the total flux Φ emitted by the source is given by this quantity multiplied by the surface area of a sphere:

$$\Phi = 4\pi^2 R^2 L. \tag{2.28}$$

At a distance r from the center of the source, this flux must uniformly irradiate a surface of area $4\pi r^2$, and thus the irradiance at distance r is given by

$$E = \frac{\pi R^2 L}{r^2}. \tag{2.29}$$

This result shows that the irradiance obeys an inverse-square law for any distance r. The intensity of this source can be determined by noting that the emitted radiation must be isotropic, and thus

$$I = \frac{\Phi}{4\pi} = \pi R^2 L. \tag{2.30}$$

If an observer at the position of dA were to look at this source, it would appear as a uniform disk of half angle

$$\theta_{1/2} = \sin^{-1}\frac{R}{r}. \tag{2.31}$$

BIBLIOGRAPHY

F. Grum and R. J. Becherer, *Optical Radiation Measurements*, Academic, New York, 1979.

M. J. Klein, *Optics*, Wiley, New York, 1970, Section 4.1.

F. E. Nicodemus, "Radiometry," in *Applied Optics and Optical Engineering*, Volume IV, R. Kingslake, ed., Academic, New York, 1965.

F. E. Nicodemus, ed., *Self Study Manual on Optical Radiation Measurements*, National Bureau of Standards Technical Note 910–1, 1976.

R.C.A. *Electro-Optics Handbook*, R.C.A., Harrison, N.J., 1974.

A. Stimson, *Photometry and Radiometry for Engineers*, Wiley, New York, 1974.

J. W. T. Walsh, *Photometry*, Dover, New York, 1958.

W. L. Wolfe, "Radiometry," in *Applied Optics and Optical Engineering*, Volume VIII, R. Kingslake, ed., Academic, New York, 1980.

C. L. Wyatt, *Radiometric Calibration: Theory and Methods*, Academic, New York, 1978.

PROBLEMS

1 Verify Eq. (2.5) of the text, and note the intuitively pleasing limiting form of this equation in the limit $\theta_{1/2} \ll 1$.

2   A disk of radius R is a non-Lambertian emitter of radiance $L_0\cos\theta$, where θ is the angle between the surface normal and the direction of observation.

(a) Calculate the radiant exitance of the source.
(b) Calculate the intensity of the source.
(c) Calculate the irradiance on a screen placed at a distance z from the center of the disk.

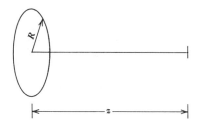

3   Describe how to build a device for measuring the radiance of a distant source. You may assume that you are given a spectrally flat detector of known responsivity.

4   A Lambertian source of radiance 10 W/m² sr is made in the shape of a ring. What is the irradiance at the point P?

$$\theta_{in} = \arcsin(0.25)$$

$$\theta_{out} = \arcsin(0.5)$$

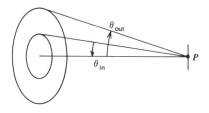

5   A spherical Lambertian scatterer of radius R is irradiated by a point source of intensity I placed at a large distance z $(z \gg R)$ from the

sphere. Find the radiance of the scattered radiation as a function of position and direction. Find an integral expression for the irradiance on the surface S resulting from scattered radiation (ignore the effect of the shadow cast by S). Evaluate this integral in the limit $r \gg R$. Discuss qualitatively how the irradiance changes as the position of S and its orientation are varied. Is the surface of the moon a Lambertian scatterer? (Please answer the last question using only your own experience, i.e., do not try to find the answer in a book.)

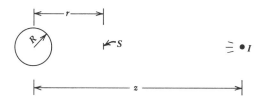

3

Theory of
Blackbody Radiation

Hot objects emit radiation that we can feel or see. Hot bodies generally become luminous at a temperature of about 500°C, and as the temperature is increased further the color of the emitted radiation changes from red to yellow to white and to blue. Common examples of objects that emit visible thermal radiation are tungsten lamps, electric heating coils, and the sun.

This chapter and the next discuss the properties of thermal radiation. The study of thermal radiation has a great deal of practical importance since many common sources of radiation are thermal sources. In addition, the theory of thermal radiation is of considerable historical interest as it provided early evidence of the necessity of energy quantization within the context of quantum mechanics. Moreover, we shall see later that noise in many detection systems arises from thermal effects, and thus the study of thermal radiation provides a basis for understanding various noise mechanisms.

We shall be interested in determining the spectral and directional properties of thermal radiation from various sources, or, using the terminology of the last chapter, in determining the spectral radiance L_ν of these sources. For an arbitrary thermal source, this is an extremely difficult problem; but, for a type of source known as a blackbody, the radiance $L_\nu(T)$ is a universal function of the optical frequency ν and the temperature T of the source.

A blackbody source can be formed by placing a small hole in the wall of an isothermal enclosure, or cavity, as shown in Fig. 3.1. If the hole is sufficiently small, it may be assumed that the radiation field within the enclosure is only slightly modified by the small amount of radiation that escapes through the hole. If the spectral radiance $L_\nu(T)$ of the radiation escaping through the hole is examined spectroscopically, it is found that $L_\nu(T)$ is a universal function that does not depend on the shape of the enclosure (as long as its dimensions are many wavelengths) or on the material used in its construction. Therefore, it can be assumed that the radiation field inside such a cavity is in thermal equilibrium with the walls of the cavity. This condition of thermal equilibrium greatly simplifies the theoretical analysis of the emitted radiation field, since the entire radiation field can be described by a single parameter, its temperature.

For the present, assume that the interior walls of the isothermal enclosure of Fig. 3.1 are at least partially absorbing, with that part of the radiation which is not absorbed being diffusely reflected. If a beam of light were to enter the cavity through the hole in the cavity wall, a negligible fraction of this radiation would be scattered in such a manner as to reemerge from the hole. Thus a blackbody source is *black* in the normal sense of the word: It absorbs all the radiation incident on it. This suggests an intimate connection between the emitting and absorbing properties of an object; this relationship is known as Kirchhoff's law and will be discussed in detail later in this chapter.

It is perhaps worth pointing out the process by which the radiation field within the cavity comes to equilibrium with the cavity walls. Any solid contains charged particles which at low temperatures are nearly stationary, but which at high temperatures are in a state of thermal agitation, either oscillating about their equilibrium positions or moving uniformly through the solid until they scatter off of some imperfection. In either case, the charged particles undergo accelerations and thereby emit radiation. The charged particles can also absorb radiation by the inverse process. Since the radiation field is thus able to exchange energy with the walls of the cavity, the two eventually come into thermal equilibrium.

3.1 RADIATION FIELD INSIDE AN ISOTHERMAL ENCLOSURE

In this section, we discuss the properties of a radiation field that is in equilibrium with the walls of an isothermal enclosure. For generality, we assume that the enclosure is filled with a uniform dielectric of refractive index N.

We first note that the radiation field does not depend on the shape or on the composition of the enclosure. Figure 3.2 shows two enclosures of different construction held at the same temperature T. They are connected

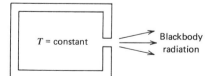

Figure 3.1. Isothermal enclosure.

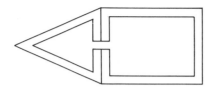

Figure 3.2. Connected isothermal enclosures.

by a small tunnel through which they can interchange radiation. If the radiation fields in the two cavities were different, more radiation could pass through the hole in one direction than the other, leading to a spontaneous cooling of one cavity and a heating of the other, in violation of the second law of thermodynamics. Since the blackbody radiation field does not depend on the shape of its confining cavity, it must be homogeneous and isotropic. Equivalently, it must be true that the radiance L is independent of position and of observation direction. These statements can be proven more explicitly (see, e.g., Reif, p. 379 of the book listed in the bibliography at the end of this chapter) by assuming that small test absorbers are placed within the enclosure at various positions and in various orientations and then by arguing that it would violate the laws of thermodynamics if these objects did not all reach the same equilibrium temperature T.

We shall next determine how the energy density, radiance, surface irradiance, and radiation pressure are related for blackbody radiation.

Relationship Between Energy Density and Radiance

In order to determine how the energy density u is related to the radiance L for blackbody radiation, we first determine the contribution to the energy density from that part of the field radiating into a given element $d\Omega$ of solid angle, and then take the sum of all such contributions to the energy density from all possible directions. The flux passing through the area element dA of Fig. 3.3a from a solid angle element $d\Omega$ inclined at angle θ to the surface

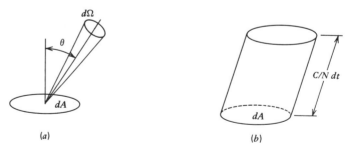

(a)   (b)

Figure 3.3. Calculation of the energy density of blackbody radiation.

normal is given by

$$d^2\Phi = L \, dA \cos \theta \, d\Omega. \tag{3.1}$$

Therefore, the energy d^3Q passing through dA in a time dt is given by this flux multiplied by dt. Since this energy is contained in a cylinder of base area dA and slant height $(c/N) \, dt$, as shown in Fig. 3.3b, the energy density associated with this radiation must be given by

$$du = \frac{d^3Q}{d^2V} = \frac{L \, dA \cos \theta \, d\Omega \, dt}{dA \cos \theta (c/N) \, dt}$$

$$= \frac{L \, d\Omega}{(c/N)}. \tag{3.2}$$

The total density within the cavity can be obtained by adding the contributions to du from all directions, giving

$$u = \int du = \frac{L}{c/N} \int_{\text{sphere}} d\Omega$$

$$= \frac{4\pi L}{(c/N)}. \tag{3.3}$$

Irradiance of Cavity Walls

Similar reasoning can be used to calculate the irradiance of the cavity wall. The contribution dE to the irradiance of the cavity wall by the flux contained in the element $d\Omega$ of solid angle inclined at an angle θ to the surface normal is given as a consequence of the definitions (2.6) and (2.3) as

$$dE = L \cos \theta \, d\Omega. \tag{3.4}$$

The total irradiance is obtained by integrating this expression over all inward directions to obtain

$$E = \int_{\substack{\text{inward} \\ \text{hemisphere}}} L \cos \theta \, d\Omega = \int_0^{\pi/2} 2\pi L \cos \theta \sin \theta \, d\theta$$

$$= \pi L. \tag{3.5}$$

Using the result (3.3), this can also be expressed as

$$E = \frac{c/N}{4} u. \tag{3.6}$$

If a small hole is drilled through the cavity wall, flux can pass out of the cavity and become available for observation. Since the radiance of the radiation field inside the cavity is independent of direction, and since it has been shown [in Eq. (2.17)] that free propagation leaves the radiance invariant, it follows that the radiance of the flux leaving the cavity is the same for all directions of observation and thus that an opening in the wall of a blackbody cavity is a Lambertian source. In particular, the radiant exitance M of such a blackbody source must equal the irradiance of the cavity walls and thus is given by

$$M = \pi L, \tag{3.7}$$

which is in agreement with the general result (2.22) for Lambertian sources.

Radiation Pressure of Blackbody Radiation

Since electromagnetic radiation carries momentum, the radiation field within a blackbody cavity will exert pressure on the walls of the cavity. We noted earlier [cf. Eq. (1.13)] that the momentum density $\mathbf{g}$ of an electromagnetic plane wave is related to its Poynting vector $\mathbf{S}$ by the relation

$$\mathbf{g} = \frac{\mathbf{S}}{(c/N)^2}. \tag{3.8}$$

From this it follows that the rate at which momentum is incident on a surface element is related to the flux Φ incident on the surface by

$$\frac{d\mathbf{p}}{dt} = \frac{\hat{\mathbf{k}}\Phi}{(c/N)}, \tag{3.9}$$

where $\hat{\mathbf{k}}$ is a unit vector in the direction of propagation. The pressure P on such a surface element is equal to the normal force $F_\perp$ per unit area, or

$$P = \frac{dF_\perp}{dA} = \frac{d^2 p_\perp}{dt\, dA}, \tag{3.10}$$

where, by Newton's second law, $dp_\perp/dt$ is equal to the normal force on the surface. Thus the contribution to the radiation pressure from the flux

$$d^2\Phi = L \, dA \cos\theta \, d\Omega$$

incident on the surface from a particular direction is given by

$$dP = \frac{d^2\Phi \cos\theta}{dA(c/N)} = \frac{L\cos^2\theta \, d\Omega}{(c/N)}. \tag{3.11}$$

The total pressure on the surface can be obtained by integrating dP over all possible directions. This integration must extend over 4π sr, since both the incoming and outgoing radiation transfer momentum to the surface and thus contribute to the pressure. The expression for the radiation pressure thus becomes

$$P = \frac{L}{(c/N)} \int_0^\pi 2\pi \cos^2\theta \sin\theta \, d\theta$$

$$= \frac{4\pi}{3}\frac{L}{c/N}, \tag{3.12}$$

which can be expressed, using Eq. (3.3), as

$$P = \tfrac{1}{3}u. \tag{3.13}$$

3.2 KIRCHHOFF'S LAW

Thus far we have restricted our attention to thermal radiation from a special class of sources known as blackbodies. Let us now consider the thermal field radiated by an arbitrary object. If such an object is placed inside an isothermal enclosure, as shown in Fig. 3.4, the object will be bathed in blackbody radiation and will eventually come to thermal equilibrium at the temperature T of the cavity walls. Once equilibrium is

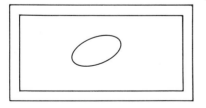

Figure 3.4. Arbitrary object within an isothermal enclosure.

achieved, the total flux Φ_e emitted by the object must equal the total flux Φ_a absorbed by the object. If the total flux incident on the object from the blackbody field is denoted by Φ_0, the absorptivity α can be defined as the fraction of the incident flux that is absorbed, or as

$$\Phi_a = \alpha\Phi_0. \tag{3.14}$$

Clearly the maximum possible value of α is unity, and an object for which $\alpha = 1$ is called a blackbody. The flux emitted by a blackbody, according to Eq. (3.14), must be equal to Φ_0, and therefore it is useful to define the emissivity ε of an arbitrary object by the relation

$$\Phi_e = \varepsilon\Phi_0. \tag{3.15}$$

The emissivity can be interpreted as the ratio of the flux emitted by a given object to the flux emitted by a blackbody of the same size, shape and temperature. Since the fluxes Φ_a and Φ_e must in thermal equilibrium be equal, we obtain Kirchhoff's law

$$\alpha = \varepsilon, \tag{3.16}$$

which implies that a good absorber of radiation is a good emitter of radiation. It is often possible to make a more precise statement of Kirchhoff's law, namely that Eq. (3.16) holds for each spectral component of the blackbody field and for each direction of the incident radiation. This result can be demonstrated by using the principle of detailed balance, and is discussed by Reif (pp. 382–386) and by Drude (pp. 496–502). This generalized form of Kirchhoff's law can, however, break down for certain materials that are not in thermal equilibrium with the radiation field and which scatter radiation with a change of direction or of wavelength (see, e.g., Landau and Lifshitz, *Statistical Physics*, p. 167).

3.3 THERMODYNAMIC PROPERTIES OF BLACKBODY RADIATION

Stefan–Boltzmann Law

The total power radiated by a hot object is a rapidly increasing function of the temperature of the object. In fact, it has been established empirically that the total power per unit area radiated by a blackbody is given by the Stefan–Boltzmann law

$$M = \sigma T^4, \tag{3.17}$$

where σ is the Stefan–Boltzmann constant whose value is given by

$$\sigma = 5.67 \times 10^{-8} \text{ W/m}^2 \text{ K}^4.$$

It is interesting that the form of the Stefan–Boltzmann law can be deduced from classical thermodynamics, although the value of the Stefan–Boltzmann constant cannot be determined using classical physics. Consider the blackbody cavity shown in Fig. 3.5, which consists of an evacuated cylinder and a tightly fitting piston. The radiation field is in equilibrium with the walls of the cylinder, which are assumed to have negligible heat capacity and are initially at temperature T. The internal energy of the system is given by

$$U = Vu(T), \tag{3.18}$$

where V is the volume containing the radiation field whose internal energy density is taken as an unknown function $u(T)$.* If an amount of heat dQ is now allowed to flow reversibly into the system, the first law of thermodynamics requires that

$$dQ = dU + dW. \tag{3.19}$$

Here dW is the work done on the piston if it is allowed to move slowly upward so as to increase the volume of the cylinder by an amount dV, and it is expressed as

$$dW = P\,dV = \tfrac{1}{3}u(T)\,dV, \tag{3.20}$$

where the relation (3.13) for the radiation pressure has been used. If this

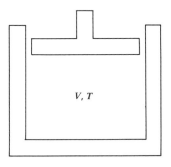

Figure 3.5. Blackbody cavity in the form of a piston.

*It is necessary to restrict the present discussion to the case of an evacuated cylinder, for which $N = 1$, so that the internal energy $u(T)$ can be taken as solely that of the radiation field.

process proceeds slowly enough that it is reversible, the entropy change due to this process is given by

$$dS = \frac{dQ}{T} = \frac{dU + dW}{T}$$

$$= \frac{u(T)dV + V\frac{du(T)}{dT}dT + \frac{1}{3}u(T)dV}{T}$$

$$= \frac{4}{3}\frac{u(T)dV}{T} + \frac{V}{T}\frac{du(T)}{dT}dT. \tag{3.21}$$

Since this process has been assumed to be reversible, it must be true that dS is an exact differential and hence that*

$$\frac{\partial^2 S}{\partial T \partial V} = \frac{\partial^2 S}{\partial V \partial T}. \tag{3.22}$$

Using Eq. (3.21), this condition implies that

$$\frac{\partial}{\partial T}\frac{4}{3}\frac{u(T)}{T} = \frac{\partial}{\partial V}\frac{V}{T}\frac{du(T)}{dT}, \tag{3.23}$$

or that

$$\frac{du(T)}{u} = 4\frac{dT}{T},$$

which has the solution

$$u(T) = aT^4, \tag{3.24}$$

where a is an undetermined constant. Finally, since the radiant exitance M,

---

*Since S must be a single-valued function of V and T, the variation in S from a change dV followed by a change dT must be the same as that resulting from a change dT followed by a change dV. Thus

$$\frac{\partial S}{\partial V}dV + \frac{\partial}{\partial T}\left(S + \frac{\partial S}{\partial V}dV\right)dT = \frac{\partial S}{\partial T}dT + \frac{\partial}{\partial V}\left(S + \frac{\partial S}{\partial T}dT\right)dV$$

or

$$\frac{\partial^2 S}{\partial T \partial V} = \frac{\partial^2 S}{\partial V \partial T}.$$

by Eqs. (3.6) and (3.7), is proportional to the energy density u, we obtain the desired result

$$M = \frac{c}{4}u(T) = \frac{c}{4}aT^4 = \sigma T^4. \tag{3.25}$$

We can likewise derive an expression for the entropy of blackbody radiation. Equation (3.21) implies that if the volume V is held constant, the entropy density $s = S/V$ changes with temperature as

$$\frac{ds}{dT} = \frac{1}{T}\frac{du(T)}{dT}, \tag{3.26}$$

which by Eq. (3.24) becomes

$$\frac{ds}{dT} = 4aT^2. \tag{3.27}$$

This equation can be integrated to give

$$s = \tfrac{4}{3}aT^3$$

$$= \frac{4}{3}\frac{u(T)}{T}. \tag{3.28}$$

Wien Displacement Law

A primary goal of the theory of blackbody radiation is to predict the spectral distribution of a blackbody radiation field. In fact, it has proven impossible to predict this distribution using only the laws of classical (i.e., non-quantum-mechanical) physics, although Wien [W. Wien, *Wied. Ann.*, **52**, 132 (1894)] was able to establish some of the properties of the spectral distribution using only these laws. The method and results obtained by Wien are outlined later; a more detailed account may be found in Richtmyer and Kennard, (see Bibliography).

We consider blackbody radiation of temperature T contained in a spherical enclosure with perfectly reflecting walls. The container is assumed to be thermally isolated and is allowed to expand slowly from initial radius R to radius $R + dR$, causing the enclosed radiation to expand adiabatically. Equation (3.19) thus becomes

$$dQ = 0 = d(uV) + \tfrac{1}{3}u\,dV$$

$$= V\,du + \tfrac{4}{3}u\,dV$$

$$= R\,dT + T\,dR, \tag{3.29}$$

where the results $u = aT^4$ and $V = \frac{4}{3}\pi R^3$ were used in obtaining the last line. This equation can now be integrated to give the result

$$RT = \text{constant}, \tag{3.30}$$

showing that the product of the radius of the enclosure and the temperature of the blackbody radiation remains constant in an adiabatic expansion.

The nature of the decrease in temperature resulting from an adiabatic expansion can alternatively be understood as follows: During the slow expansion of the sphere, the enclosed radiation is reflected many times from the reflecting surface, which is moving slowly away from it. Reflection from a moving mirror leads to a Doppler shift in the reflected radiation, and the effect of this Doppler shift is to increase the wavelength of any spectral component of the radiation field in direct proportion to the increase in the radius of the sphere, so that

$$\frac{\lambda}{R} = \text{constant}. \tag{3.31}$$

When this result is combined with that of Eq. (3.30), we obtain the result

$$\lambda T = \text{constant}. \tag{3.32}$$

Let us now restrict our attention to the energy density $u_\lambda \Delta\lambda$ contained in a small spectral interval $\Delta\lambda$ centered on the wavelength λ. Since we can assume that distinct spectral bands do not interact with one another, Eq. (3.19) can be applied to this spectral interval alone, leading to

$$dQ = 0 = d(u_\lambda \Delta\lambda V) + \frac{1}{3}u_\lambda \Delta\lambda \, dV$$

$$= u_\lambda V d(\Delta\lambda) + V\Delta\lambda \, du_\lambda + \frac{4}{3}u_\lambda \Delta\lambda \, dV \tag{3.33}$$

or

$$\frac{d(\Delta\lambda)}{\Delta\lambda} + \frac{du\,\lambda}{u_\lambda} + \frac{4}{3}\frac{dV}{V} = 0. \tag{3.34}$$

This expression can be simplified by noting that $d(\Delta\lambda)/\Delta\lambda = d\lambda/\lambda$ and that $dV/V = 3\,dR/R = 3\,d\lambda/\lambda$, giving

$$\frac{5\,d\lambda}{\lambda} + \frac{du_\lambda}{u_\lambda} = 0, \tag{3.35}$$

which can be integrated to obtain

$$\lambda^5 u_\lambda = \text{constant}. \tag{3.36}$$

This result can now be combined with that of Eq. (3.32) to conclude that

$$u_\lambda = \lambda^{-5} f(\lambda T) \tag{3.37}$$

where $f(\lambda T)$ is some undetermined function of the product λT. This result entails Wien's displacement law, which states that the shape of the blackbody spectral distribution function depends on the temperature T only as the product λT.

3.4 PLANCK RADIATION LAW

The correct theoretical expression for the spectral distribution of blackbody radiation was discovered by Planck in 1900. Planck found that in order to obtain an analytic expression that agreed with experimental results, he had to assume that the energy of the oscillators with which he modeled the radiation field was not a continuously varying quantity but was quantized in units of $\hbar\omega$. This was a revolutionary concept at the time and provided evidence of the quantum mechanical features of nature.

In this section, we derive the Planck radiation law by first decomposing the blackbody radiation field into its normal modes, then determining the average energy per mode through use of the laws of statistical mechanics, and finally determining the spectral energy density u_ω of the radiation field by multiplying the expression for the density of modes per unit frequency interval by the expression for the average energy per mode. This method was introduced by Rayleigh (1900) and Jeans (1905), who calculated the average energy per mode using the laws of classical physics and thus obtained a form of the radiation law that is valid only in the limiting case where quantum mechanical features can be neglected. The calculation presented here properly treats these quantum mechanical features.

The Planck radiation law can also be derived by treating the radiation field as an ideal gas of massless particles that obey Bose–Einstein statistics. This derivation will not be given here, but can be found in R. H. Fowler (*Statistical Mechanics*, Macmillan, New York, 1936, pp. 117–118), or in R. B. Leighton (*Principles of Modern Physics*, McGraw-Hill, New York, 1959, pp. 365–367).

Density of Field Modes

For any finite region of space and any finite spectral interval, it is possible to describe the electromagnetic field in terms of a finite number of degrees of freedom, which can be taken as the amplitudes of the normal modes of the field for the volume under consideration. It has been shown [H. Weyl, *Math. Ann.*, **71**, 441 (1911); R. Courant, *Math. Z.*, **7**, 14 (1920)] that, as long as the characteristic dimensions of this region are much greater than the wavelength of the radiation being considered, the density of modes is independent of the exact size or shape of this volume. In fact, this is very reasonable, since the density of modes is a property of the radiation field and not of the boundary surface that encloses it. Thus we calculate the mode density for a particularly simple enclosure, a cube of side L with perfectly (electrically) conducting walls, filled with a non-dispersive dielectric of refractive index N, as shown in Figure 3.6. Mathematically, the problem is one of finding all of the possible solutions to the wave equation

$$\nabla^2 \psi = \frac{1}{(c/N)^2} \frac{\partial^2 \psi}{\partial t^2}, \tag{3.38}$$

ψ being any Cartesian component of $\mathbf{E}(\mathbf{r}, t)$ or $\mathbf{H}(\mathbf{r}, t)$, where the solutions are required to meet boundary conditions at the cavity walls appropriate to a metallic surface [cf. the discussion following Eq. (1.34)] and to satisfy Maxwell's equations. The solution for $\mathbf{E}(\mathbf{r}, t)$ for our boundary conditions has the form (cf. Problem 1.6)

$$E_x(\mathbf{r}, t) = E_x e^{-i\omega t}\cos(k_x x)\sin(k_y y)\sin(k_z z), \tag{3.39a}$$

$$E_y(\mathbf{r}, t) = E_y e^{-i\omega t}\sin(k_x x)\cos(k_y y)\sin(k_z z), \tag{3.39b}$$

$$E_z(\mathbf{r}, t) = E_z e^{-i\omega t}\sin(k_x x)\sin(k_y y)\cos(k_z z). \tag{3.39c}$$

Here the components of the wave vector $\mathbf{k}$ must satisfy

$$k_x^2 + k_y^2 + k_z^2 = \frac{\omega^2}{(c/N)^2}, \tag{3.40}$$

and the boundary conditions require that the components of $\mathbf{k}$ be integral

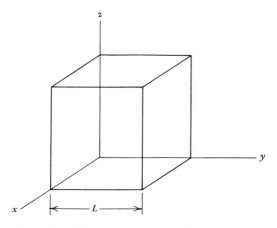

Figure 3.6.   Reference volume for mode decomposition.

multiples of π/L, that is,

$$k_x = n_x \frac{\pi}{L} \tag{3.41a}$$

$$k_y = n_y \frac{\pi}{L} \tag{3.41b}$$

$$k_z = n_z \frac{\pi}{L}, \tag{3.41c}$$

where n_x, n_y, n_z are nonnegative integers, no more than one of which can be zero. As in the case of freely propagating plane waves, the Maxwell equation $\nabla \cdot \mathbf{E} = 0$ leads to the constraint

$$\mathbf{k} \cdot \mathbf{E} = 0, \tag{3.42}$$

implying that two independent solutions, and thus two modes of the field, exist for each allowed value of the wave vector.

The resulting density of field modes can be calculated with the aid of Fig. 3.7, in which each allowed value of the wave vector is shown as a lattice point in a three-dimensional $\mathbf{k}$ space. The number of field modes for which the magnitude of the wave vector lies between k and $k + dk$ is given by the volume of the corresponding region in $\mathbf{k}$ space divided by the volume $(\pi/L)^3$ surrounding each lattice point, leading to the expression

$$\frac{\frac{1}{8}\left(4\pi k^2 \, dk\right)}{(\pi/L)^3} \times 2.$$

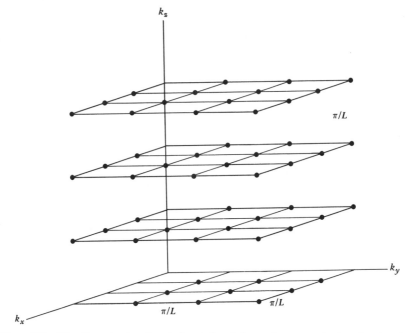

Figure 3.7. The allowed values $\mathbf{k} = k_x\hat{\mathbf{x}} + k_y\hat{\mathbf{y}} + k_z\hat{\mathbf{z}}$ are shown as a lattice of points in k space.

The factor $\frac{1}{8}$ accounts for the fact that, since it has been assumed that all of the components of $\mathbf{k}$ are positive, the appropriate volume element consists of only one octant of a shell of radius k and thickness dk, while the final factor of 2 accounts for the two possible polarizations allowed for each value of $\mathbf{k}$. If $\rho_k\, dk$ is defined as the number of modes per unit volume having the magnitude of their wave vector in the range k to $k + dk$, this quantity may be expressed by

$$\rho_k\, dk = \frac{k^2\, dk}{\pi^2}. \tag{3.43}$$

Finally, using the relation $\omega = k(c/N)$, we obtain the desired result for the quantity $\rho_\omega\, d\omega$, defined as the number of modes per unit volume whose angular frequencies lie in the range ω to $\omega + d\omega$:

$$\rho_\omega\, d\omega = \frac{\omega^2\, d\omega}{\pi^2(c/N)^3}. \tag{3.44}$$

For purposes of reference, we also quote the mode density for linear frequency units ($\nu = \omega/2\pi$)

$$\rho_\nu \, d\nu = \frac{8\pi\nu^2 \, d\nu}{(c/N)^3},$$ (3.45)

and for wavelength units [$\lambda = (c/N)/\nu$]

$$\rho_\lambda \, d\lambda = \frac{8\pi}{\lambda^4} d\lambda.$$ (3.46)

Calculation of the Energy Per Mode

From the discussion in the preceding paragraphs, we see that the radiation field in the volume under consideration can be formed as a superposition of solutions of the form (3.40), where the field strength **E** corresponding to any normal mode oscillates harmonically in time. This fact suggests that we treat the electromagnetic field as a collection of harmonic oscillators, each mode of the field being represented by a single oscillator. If each oscillator is supposed to be in thermal equilibrium with the walls of the cavity, the average energy of the oscillator can be calculated through the use of statistical mechanics. An oscillator of resonance frequency ω can be characterized by a coordinate q and a momentum p, such that the energy of the oscillator is given by

$$E(p,q) = \frac{p^2}{2m} + \tfrac{1}{2}m\omega^2 q^2.$$ (3.47)

If it is assumed that the energy of the oscillator is distributed according to the laws of classical statistical mechanics, the mean energy of the oscillator is given by

$$\bar{E} = \frac{\iint E(p,q)e^{-E(p,q)/kT} \, dp \, dq}{\iint e^{-E(p,q)/kT} \, dp \, dq}.$$ (3.48)

The integration can be simply performed for the energy function $E(p,q)$ of Eq. (3.47) to give

$$\bar{E} = kT.$$ (3.49)

This result is in accordance with the equipartion theorem which states that the mean energy per degree of freedom for a classical system is equal to

$\frac{1}{2}kT$. Combining this result with expression (3.44) for the density of field modes, we obtain the Rayleigh–Jeans law

$$u_\omega = \frac{\omega^2}{\pi^2(c/N)^3}kT, \qquad (3.50)$$

which predicts the spectral energy density of the blackbody field. This radiation distribution was derived by Lord Rayleigh, who inadvertently omitted the factor of $\frac{1}{8}$ mentioned in the derivation of Eq. (3.43) [*Philos. Mag.*, **49**, 539 (1900)], and later by Jeans, who corrected this error [*Philos. Mag.* **10**, 91 (1905)]. The Rayleigh–Jeans law is of great historical interest, in part because most subsequent theories of blackbody radiation have been based on the concept of a mode decomposition of the radiation field. However, the Rayleigh–Jeans law clearly cannot be valid at large frequencies, since Eq. (3.50) predicts that u_ω diverges as ω^2 for large ω. This unphysical consequence is known as the ultraviolet catastrophe.

The nature of the error in the classical argument just presented is in the assumption that the energies of a classical harmonic oscillator are continuously distributed. Planck [*Verh. Dtsch. Phys. Ges.* **2**, 202 (1900)] made the assumption that the permitted energies of a harmonic oscillator are of the form

$$E = n\hbar\omega, \qquad n = 0, 1, 2, \ldots, \qquad (3.51)$$

and on the basis of this assumption was able to deduce a radiation law that is in accord with experience. In fact, a fully quantum mechanical treatment of the free electromagnetic field (see, e.g., R. Loudon, *The Quantum Theory of Light*, Oxford University, London 1973) shows that the energy of each mode of the field is quantized in units of $\hbar\omega$; these quanta have become known as photons.

Let us consider the probability distribution function for the integer n of Eq. (3.51), which can be interpreted as the photon occupation number, that is, the number of photons per mode. The probability that n photons are excited in a mode of angular frequency ω is given by

$$p(n) = \frac{e^{-n\hbar\omega/kT}}{\sum_{n=0}^{\infty}e^{-n\hbar\omega/kT}}. \qquad (3.52)$$

Here the numerator is the Boltzmann factor and the denominator ensures that the probability is properly normalized. The denominator has the form of a geometrical series and is equal to $(1 - e^{-\hbar\omega/kT})^{-1}$; thus the probability distribution (3.52) can be expressed as

$$p(n) = (1 - e^{-\hbar\omega/kT})e^{-n\hbar\omega/kT}. \tag{3.53}$$

The average value of the occupation number is given in general by

$$\bar{n} = \sum_{n=0}^{\infty} np(n). \tag{3.54}$$

This expression can be evaluated most readily by defining $x = \hbar\omega/kT$ and using Eqs. (3.53) in (3.54) to give

$$\bar{n} = (1 - e^{-x}) \sum_{n=0}^{\infty} ne^{-nx}$$

$$= -(1 - e^{-x})\frac{d}{dx} \sum_{n=0}^{\infty} e^{-nx}$$

$$= -(1 - e^{-x})\frac{d}{dx}\left(\frac{1}{1 - e^{-x}}\right)$$

$$= \frac{e^{-x}}{1 - e^{-x}} = \frac{1}{e^x - 1}$$

or

$$\bar{n} = \frac{1}{e^{\hbar\omega/kT} - 1}. \tag{3.55}$$

This result, known as the Planck distribution, is illustrated graphically in Fig. 3.8. It is useful to note the two limiting forms of the Planck distribution. In the high-temperature limit, Eq. (3.55) becomes

$$\bar{n} = \frac{kT}{\hbar\omega} \quad \text{for } \hbar\omega \ll kT, \tag{3.56}$$

and thus by (3.51)

$$\bar{E} = kT \quad \text{for } \hbar\omega \ll kT,$$

in accordance with the result (3.49) predicted by the classical equipartition theorem. We note that by Eq. (3.56), the limit $\hbar\omega \ll kT$ necessarily implies that $\bar{n} \gg 1$; therefore, the classical result is expected to hold as a conse-

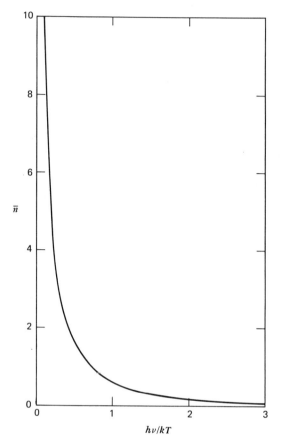

Figure 3.8. Planck distribution, giving the mean number of photons per mode of a thermal radiation field.

quence of the correspondence principle. In the other limiting case, we obtain

$$\bar{n} = e^{-\hbar\omega/kT} \quad \text{for } \hbar\omega \gg kT. \tag{3.57}$$

For this limit, we have $\bar{n} \ll 1$, and the average energy per mode is given by

$$\bar{E} = \hbar\omega e^{-\hbar\omega/kT} \quad \text{for } \hbar\omega \gg kT.$$

By eliminating the quantity $x = \hbar\omega/kT$ from Eqs. (3.53) and (3.55), it is possible to rewrite the photon probability distribution as

$$p(n) = \frac{\bar{n}^n}{(1 + \bar{n})^{1+n}}, \tag{3.58}$$

which is known as the Bose–Einstein probability distribution.

Since there is a distribution in the number of photons excited per mode, it is useful to investigate the root-mean-square dispersion in this number, which is defined as

$$(\Delta n)_{\text{rms}} \equiv \sqrt{\overline{(n - \bar{n})^2}} . \tag{3.59}$$

This expression can be simplified as follows:

$$(\Delta n)_{\text{rms}} = \sqrt{\overline{n^2 - 2n\bar{n} + \bar{n}^2}}$$

$$= \sqrt{\overline{n^2} - 2\bar{n}^2 + \bar{n}^2}$$

$$= \sqrt{\overline{n^2} - \bar{n}^2} . \tag{3.60}$$

The quantity $\overline{n^2}$ can be calculated by a procedure similar to that used to calculate $\bar{n}$. Again letting $x = \hbar\omega/kT$, we obtain

$$\overline{n^2} = \sum_{n=0}^{\infty} n^2 p(n)$$

$$= (1 - e^{-x}) \frac{d^2}{dx^2} \sum_{n=0}^{\infty} e^{-nx}$$

$$= (1 - e^{-x}) \frac{d^2}{dx^2} \left(\frac{1}{1 - e^{-x}} \right)$$

$$= \frac{e^x + 1}{(e^x - 1)^2} .$$

By recalling that $\bar{n} = (e^x - 1)^{-1}$, this result may be expressed as

$$\overline{n^2} = 2\bar{n}^2 + \bar{n}, \tag{3.61}$$

and thus the expression (3.60) for the rms dispersion becomes

$$(\Delta n)_{\text{rms}} = \sqrt{\bar{n}(\bar{n} + 1)} . \tag{3.62}$$

It is again worth noting the two limiting forms of this expression. For $\bar{n} \ll 1$, corresponding to the case $\hbar\omega \gg kT$, the rms dispersion is given by

$$(\Delta n)_{\text{rms}} \simeq \sqrt{\bar{n}} , \tag{3.63}$$

which is typical of the statistical behavior shown by systems composed of discrete, statistically independent particles and thus reflects the particulate nature of the quantized electromagnetic field in this limit. On the other hand, the limit $\bar{n} \gg 1$, corresponding to $\hbar\omega \ll kT$, implies a dispersion given by

$$(\Delta n)_{\text{rms}} \simeq \bar{n}. \tag{3.64}$$

The increased dispersion in this case results from the fact that photons obey Bose–Einstein statistics and thus tend to "bunch." It has been shown (see, e.g., D. ter Haar, in *Quantum Optics*, R. J. Glauber, ed., Academic, New York, 1969) that dispersion of this sort is expected from the stochastic interference of waves and thus reflects the wavelike properties of the field at large photon occupation numbers. More will be said later (cf., especially, Chapter 14) about these fluctuations in the photon occupation number in connection with the discussion of noise in the detection process.

The correct form for the spectrum of blackbody radiation can be obtained by combining expression (3.45) for the density of modes with expressions (3.51) and (3.55), whose product gives the average energy per mode, to obtain

$$u_\nu = \rho_\nu \bar{n} h\nu$$

$$= \frac{8\pi}{(c/N)^3} \frac{h\nu^3}{e^{h\nu/kT} - 1}. \tag{3.65}$$

The result is known as the Planck radiation law and gives the energy per unit volume per unit frequency interval for blackbody radiation.

In our earlier discussion of the Stefan–Boltzmann law, it was noted that the form of this law, but not the value of the Stefan–Boltzmann constant, could be obtained by means of thermodynamic reasoning. The value of this constant can now be obtained through use of the Planck radiation law. The total energy density of blackbody radiation must be given by the integral over all frequencies of the quantity u_ν of Eq. (3.65), giving

$$u = \int_0^\infty u_\nu \, d\nu. \tag{3.66}$$

By making the substitution $x = h\nu/kT$, this integral may be expressed as

$$u = \frac{8\pi k^4 T^4}{(c/N)^3 h^3} \int_0^\infty \frac{x^3 \, dx}{e^x - 1}. \tag{3.67}$$

The definite integral in this expression has the value $\pi^4/15$. The total flux per unit area emitted by a blackbody can then be obtained through use of relation (3.6), giving

$$M = \frac{1}{4}\frac{c}{N}u$$

$$= N^2\left(\frac{2\pi^5 k^4}{15h^3 c^2}\right)T^4. \tag{3.68}$$

The quantity in parentheses can thus be identified as the Stefan–Boltzmann constant σ, whose value is 5.67×10^{-8} W/m² K⁴.

The spectral radiance of blackbody radiation can be obtained from Eq. (3.65) for the spectral energy density and the general relation (3.3) between these quantities and is expressed as

$$L_\nu = \frac{(c/N)}{4\pi}u_\nu$$

$$= \frac{2h\nu^3}{(c/N)^2}\frac{1}{e^{h\nu/kT} - 1}. \tag{3.69}$$

This result is also known as the Planck radiation law.

BIBLIOGRAPHY

P. Drude, *The Theory of Optics*, Dover, New York, 1959, Part III.

N. H. Frank, *Introduction to Electricity and Optics*, McGraw-Hill, New York, 1950, Chapter 20.

D. ter Haar, in *Quantum Optics*, R. J. Glauber, ed., Academic, New York, 1969.

A. L. King, *Thermophysics*, Freeman, San Francisco, 1962, Chapter 16.

L. D. Landau and E. M. Lifshitz, *Statistical Physics*, Addison-Wesley, Reading, Mass., 1969, Section 60.

R. B. Leighton, *Principles of Modern Physics*, McGraw-Hill, New York, 1959, Section 2.1.

R. Loudon, *The Quantum Theory of Light*, Oxford University, London, 1973, Chapter 1.

F. Reif, *Fundamentals of Statistical and Thermal Physics*, McGraw-Hill, New York, 1965, Chapter 9.

F. K. Richtmyer and E. H. Kennard, *Introduction to Modern Physics*, McGraw-Hill, New York, 1947, Chapter 5.

A. Sommerfeld, *Thermodynamics and Statistical Mechanics*, Academic, New York, 1964, Section 20.

PROBLEMS

1 Determine numerically the mean value of the photon occupation number for sunlight at wavelengths of 0.5, 1.0, 10, and 100 μm. Assume the sun is a blackbody emitter of temperature 5800 K.

   For the same wavelengths, determine the mean value of the photon occupation number for a room-temperature blackbody.

2 What is the radiation pressure of blackbody radiation of temperature 5800 K?

   At what temperature does the pressure of blackbody radiation equal 1 atm?

3 Assuming that the sun is a blackbody of temperature 5800 K, and neglecting absorption in the earth's atmosphere, what is the flux incident on a solar collector of area 1 m^2 whose normal points in the direction of the sun?

4 The main heat load to the liquid-helium cryostat shown below is room-temperature blackbody radiation entering through a window of area 1 cm^2. How much power is dissipated in the liquid helium? At what rate does the helium boil away?

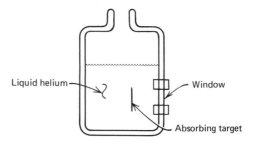

5 How many photons are contained in a unit volume of blackbody radiation of temperature T? Evaluate this expression to determine how many photons are contained in 1 cm^3 at room temperature.

   What is the total photon irradiance of a surface exposed to blackbody radiation of temperature T? Evaluate this expression to determine the photon flux incident on a 1-cm^2 area exposed to room-temperature blackbody radiation.

4

Applications of
Blackbody Radiation Theory

This chapter discusses several topics involving the application of the theory of blackbody radiation to problems of a somewhat practical nature. The Planck radiation law, derived in the preceding chapter, is discussed here with particular emphasis placed on numerical evaluation of the radiometric quantities. A phenomenological description of thermal sources that are not blackbodies is presented next. Finally, two topics from the field of radiative heat transfer are presented. One topic entails calculating the rate at which energy is transferred in vacuum between nonblack objects by thermal radiation, while the other involves treating the propagation of thermal radiation through absorbing media.

4.1 CALCULATIONS INVOLVING THE PLANCK RADIATION LAW

In Chapter 3, it was shown [cf. Eq. (3.69)] that the spectral radiance (in frequency units) of blackbody radiation contained in a region of unit refractive index is given by the expression

$$L_\nu = \frac{2h\nu^3}{c^2} \frac{1}{e^{h\nu/kT} - 1}.$$ (4.1)

This result, known as the Planck radiation law, is shown in Fig. 4.1, where it has been normalized such that its maximum value is unity. Also shown here is the Rayleigh–Jeans radiation law [cf. Eq. (3.50)] and the Wien radiation law, which is given by Eq. (4.1) with the factor $e^{h\nu/kT} - 1$ in the denominator replaced by $e^{h\nu/kT}$. The Wien law is thus accurate in the limit $h\nu \gg kT$, while the Rayleigh–Jeans law is accurate in the opposite limit $h\nu \ll kT$.

The frequency at which the function L_ν reaches its maximum value can be determined by setting the derivative with respect to ν of Eq. (4.1) equal to

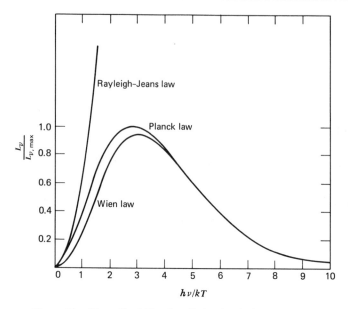

Figure 4.1. Normalized Planck radiation law in frequency units.

zero. Again letting $x = h\nu/kT$, this condition becomes

$$e^{-\lambda} = 1 - \tfrac{1}{3}x. \tag{4.2}$$

This is a transcendental equation; it can be solved numerically to give the value $x \simeq 2.821$. Thus the frequency at which L_ν is a maximum is given by

$$\nu_{L_\nu(\mathrm{max})} = 2.821\frac{kT}{h}, \tag{4.3}$$

and, since $\lambda = c/\nu$, the corresponding wavelength is given by

$$\lambda_{L_\nu(\mathrm{max})}T = \frac{ch}{2.821k}. \tag{4.4}$$

This result illustrates Wien's displacement law since the product of the temperature and the wavelength corresponding to the peak in L_ν is a constant. This expression can be numerically evaluated to obtain

$$\lambda_{L_\nu(\mathrm{max})}T = 5.100 \times 10^{-3}\,(\mathrm{m\ K}). \tag{4.5}$$

Thus, for instance, L_ν for a blackbody at the temperature $T = 5800$ K (roughly that of the surface of the sun) peaks at a wavelength of 0.879 μm in vacuum.

For certain computational purposes, it is useful to express the spectral radiance L_ν as the product

$$L_\nu = L_{\nu,\mathrm{max}}(T) f_\nu\left(\frac{h\nu}{kT}\right),$$

where $L_{\nu,\mathrm{max}}(T)$ is the maximum value of the spectral radiance of a blackbody of temperature T and where $f_\nu(h\nu/kT)$ is the spectral distribution of the Planck radiation law in frequency units, normalized such that its maximum value is unity. In order to obtain this form, we first rewrite Eq. (4.1) as

$$L_\nu = \left[\frac{2k^3T^3}{c^2h^2}\right]\left[\frac{(h\nu/kT)^3}{e^{h\nu/kT} - 1}\right], \tag{4.6}$$

where the first factor is independent of ν and where the second factor depends on ν only as the combination $h\nu/kT$. The value of the frequency that maximizes the second factor is given as $\nu = 2.821kT/h$ according to Eq. (4.3), and at this frequency the second factor has the value 1.421. Thus we can identify the normalized spectral distribution function as

$$f_\nu\left(\frac{h\nu}{kT}\right) = \frac{1}{1.421}\frac{(h\nu/kT)^3}{e^{h\nu/kT} - 1}, \tag{4.7}$$

which is the functional form plotted in Fig. 4.1, and we can identify the maximum value of the spectral radiance as

$$L_{\nu,\mathrm{max}}(T) = \frac{2.842k^3T^3}{c^2h^2}. \tag{4.8}$$

This expression can be evaluated numerically to give

$$L_{\nu,\mathrm{max}}(T) = 1.896 \times 10^{-19}T^3\left(\frac{1}{\mathrm{K}^3}\right)\left(\frac{\mathrm{W}}{\mathrm{m}^2\,\mathrm{Hz}\,\mathrm{sr}}\right). \tag{4.9}$$

In many practical applications, it is more convenient to work with the radiance per unit wavelength interval L_λ rather than the quantity L_ν. These two quantities are related by

$$L_\lambda|\,d\lambda| = L_\nu|\,d\nu| \tag{4.10}$$

where

$$\lambda = \frac{c}{\nu}. \tag{4.11}$$

Expression (4.1) for the spectral radiance in frequency units can thus be used to obtain the relation

$$L_\lambda = \frac{2hc^2}{\lambda^5} \frac{1}{e^{hc/\lambda kT} - 1} \tag{4.12}$$

which is the Planck radiation law in wavelength units. The quantity appearing in the exponent is equal to the previously introduced parameter x, which is equivalently expressed as

$$x = \frac{h\nu}{kT} = \frac{hc}{\lambda kT}. \tag{4.13}$$

This parameter can be evaluated numerically as

$$x = \frac{0.01439}{\lambda T} \text{ (m K)} = \frac{1.439}{\lambda T} \text{ (cm K)} = \frac{1.439 \times 10^4}{\lambda T} \text{ (}\mu\text{m K)}.$$

$$\tag{4.14}$$

A normalized form of the spectral distribution of Eq. (4.12) is shown in Fig. 4.2. The wavelength at which the function L_λ achieves its maximum value is obtained by setting the derivative with respect to λ of Eq. (4.12) equal to zero, requiring that

$$e^{-x} = 1 - \tfrac{1}{5}x, \tag{4.15}$$

which can be solved numerically to give $x \simeq 4.965$. The wavelength at which L_λ is a maximum is thus given by

$$\lambda_{L_\lambda(\text{max})}T = \frac{1}{4.965} \frac{hc}{k}, \tag{4.16}$$

which can be evaluated numerically to give

$$\lambda_{L_\lambda(\text{max})}T = 2.898 \times 10^{-3} \text{ (m K)}. \tag{4.17}$$

As an example, L_λ for a blackbody at the temperature $T = 5800$ K, peaks at

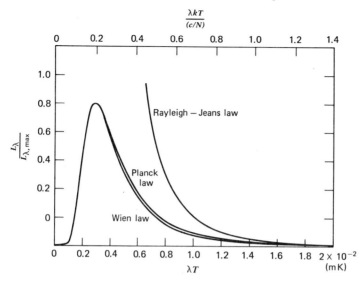

Figure 4.2. Normalized Planck radiation law in wavelength units.

a wavelength of 0.500 μm. By comparison with Eq. (4.5), we note that the functions L_ν and L_λ peak at different wavelengths.

It can be useful to express the relation for the spectral radiance L_λ as the product of its maximum value at a fixed temperature $L_{\lambda,\max}(T)$ and a normalized spectral distribution function $f_\lambda(\lambda kT/hc)$ for the Planck radiation law in wavelength units, leading to an expression of the form

$$L_\lambda = L_{\lambda,\max}(T) f_\lambda\left(\frac{\lambda kT}{hc}\right). \qquad (4.18)$$

To establish this result, Eq. (4.12) is expressed as

$$L_\lambda = \frac{2k^5 T^5}{h^4 c^3} \frac{(hc)/(\lambda kT)^5}{e^{hc/\lambda kT} - 1}. \qquad (4.19)$$

Since the second factor reaches its maximum value of 21.20 at $x = 4.965$, it is possible to identify the maximum value of L_λ as

$$L_{\lambda,\max}(T) = \frac{42.40 k^5 T^5}{h^4 c^3}, \qquad (4.20)$$

Table 4.1. Normalized Values of the Spectral Radiance L_λ, Integrated Spectral Radiance $\int L_\lambda\, d\lambda$, Spectral Photon Radiance $L_{p,\lambda}$, Integrated Spectral Photon Radiance $\int L_{p,\lambda}\, d\lambda$, and Spectral Radiance in Frequency Units L_ν

λT	$x = \dfrac{h\nu}{kT}$	$\dfrac{\int_0^\lambda L_\lambda\, d\lambda}{\int_0^\infty L_\lambda\, d\lambda}$	$\dfrac{L_\lambda}{L_{\lambda,\mathrm{max}}}$	$\dfrac{\int_0^\lambda L_{p,\lambda}\, d\lambda}{\int_0^\infty L_{p,\lambda}\, d\lambda}$	$\dfrac{L_{p,\lambda}}{L_{p,\lambda,\mathrm{max}}}$	$\dfrac{L_\nu}{L_{\nu,\mathrm{max}}}$
cm K	large x	$\dfrac{x^3 e^{-x}}{6.4939}$	$\dfrac{x^5 e^{-x}}{21.201}$	$\dfrac{x^2 e^{-x}}{2.404}$	$\dfrac{x^4 e^{-x}}{4.780}$	$\dfrac{x^3 e^{-x}}{1.4214}$
0.00	↑	↑	↑	↑	↑	↑
0.01	143.883	$0.0^{56}16$	$0.0^{53}95$	$0.0^{58}31$	$0.0^{54}29$	$0.0^{56}68$
0.02	71.942	$0.0^{26}37$	$0.0^{23}52$	$0.0^{27}14$	$0.0^{24}32$	$0.0^{25}15$
0.03	47.961	$0.0^{16}27$	$0.0^{13}18$	$0.0^{17}15$	$0.0^{14}16$	$0.0^{15}12$
0.04	35.971	$0.0^{11}19$	$0.0^9 678$	$0.0^{12}14$	$0.0^{10}84$	$0.0^{11}78$
0.05	28.777	$0.0^8 130$	$0.0^6 296$	$0.0^9 117$	$0.0^7 456$	$0.0^8 533$
0.055	26.161	$0.0^7 135$	$0.0^5 251$	$0.0^8 134$	$0.0^6 426$	$0.0^7 548$
0.06	23.980	$0.0^7 929$	$0.0^4 144$	$0.0^7 100$	$0.0^5 266$	$0.0^6 373$
0.065	22.136	$0.0^6 467$	$0.0^4 610$	$0.0^7 543$	$0.0^4 122$	$0.0^5 186$
0.07	20.555	$0.0^5 184$	$0.0^3 205$	$0.0^6 220$	$0.0^4 442$	$0.0^5 723$
0.075	19.184	$0.0^5 594$	$0.0^3 571$	$0.0^6 791$	$0.0^3 132$	$0.0^4 231$
0.08	17.985	$0.0^4 164$	0.00137	$0.0^5 232$	$0.0^3 338$	$0.0^4 633$
0.085	16.927	$0.0^4 399$	0.00292	$0.0^5 597$	$0.0^3 765$	$0.0^3 152$
0.09	15.987	$0.0^4 870$	0.00562	$0.0^4 137$	0.00156	$0.0^3 328$
0.095	15.146	$0.0^3 173$	0.00994	$0.0^4 288$	0.00291	$0.0^3 646$
0.10	14.388	$0.0^3 321$	0.01640	$0.0^4 558$	0.00506	0.00118
0.11	13.080	$0.0^3 911$	0.03767	$0.0^3 173$	0.01278	0.00328
0.12	11.990	0.00213	0.07253	$0.0^3 438$	0.02684	0.00752
0.13	11.068	0.00432	0.12225	$0.0^3 951$	0.04898	0.01488
0.14	10.277	0.00779	0.18606	0.00183	0.08030	0.02628
0.15	9.592	0.01285	0.26147	0.00321	0.12091	0.04239
0.16	8.993	0.01971	0.34488	0.00522	0.17011	0.06361
0.17	8.464	0.02853	0.43231	0.00795	0.22656	0.09001
0.18	7.994	0.03933	0.51993	0.01150	0.28851	0.12137
0.19	7.573	0.05210	0.60440	0.01594	0.35402	0.15720
0.20	7.194	0.06672	0.68310	0.02129	0.42117	0.19686
0.22	6.540	0.10087	0.81632	0.03478	0.55363	0.28467
0.24	5.995	0.14024	0.91215	0.05179	0.67487	0.37854
0.26	5.534	0.18310	0.97090	0.07192	0.77819	0.47286
0.28	5.139	0.22787	0.99713	0.09461	0.86070	0.56323
0.30	4.796	0.27320	0.99717	0.11930	0.92220	0.64658
0.32	4.496	0.31807	0.97740	0.14541	0.96420	0.72110
0.34	4.232	0.36170	0.94358	0.17243	0.98901	0.78587
0.36	3.997	0.40327	0.90046	0.19994	0.99933	0.84078
0.38	3.786	0.44334	0.85177	0.22756	0.99781	0.88615
0.40	3.597	0.48084	0.80032	0.25500	0.98686	0.92258
0.45	3.197	0.56428	0.67164	0.32147	0.93174	0.97990

Table 4.1. (*Continued*)

λT	$x = \dfrac{h\nu}{kT}$	$\dfrac{\int_0^\lambda L_\lambda \, d\lambda}{\int_0^\infty L_\lambda \, d\lambda}$	$\dfrac{L_\lambda}{L_{\lambda,\max}}$	$\dfrac{\int_0^\lambda L_{p,\lambda} \, d\lambda}{\int_0^\infty L_{p,\lambda} \, d\lambda}$	$\dfrac{L_{p,\lambda}}{L_{p,\lambda,\max}}$	$\dfrac{L_\nu}{L_{\nu,\max}}$
0.50	2.878	0.63370	0.55493	0.38328	0.85534	0.99951
0.55	2.616	0.69086	0.45572	0.43953	0.77269	0.99321
0.60	2.398	0.73777	0.37399	0.49009	0.69175	0.97001
0.65	2.214	0.77630	0.30764	0.53525	0.61645	0.93645
0.7	2.0555	0.80806	0.25411	0.57542	0.54835	0.89708
0.8	1.7985	0.85624	0.17610	0.64299	0.43428	0.81196
0.9	1.5987	0.88998	0.12481	0.69665	0.34629	0.72838
1.0	1.4388	0.91415	0.09045	0.73963	0.27883	0.65166
1.1	1.3080	0.93184	0.06692	0.77442	0.22692	0.58337
1.2	1.1990	0.94505	0.05045	0.80287	0.18664	0.52343
1.3	1.1068	0.95509	0.03869	0.82640	0.15506	0.47112
1.4	1.0277	0.96285	0.03013	0.84603	0.13005	0.42552
1.5	0.9592	0.96893	0.02380	0.86257	0.11004	0.38574
1.6	0.8993	0.97376	0.01903	0.87662	0.09386	0.35095
1.7	0.8464	0.97765	0.01539	0.88864	0.08065	0.32042
1.8	0.7994	0.98081	0.01258	0.89901	0.06978	0.29354
1.9	0.7573	0.98340	0.01037	0.90801	0.06076	0.26979
2.0	0.7194	0.98555	0.00863	0.91587	0.05321	0.24871
2.5	0.5755	0.99216	0.00383	0.94339	0.02950	0.17237
3.0	0.4796	0.99529	0.00194	0.95936	0.01799	0.12611
3.5	0.4111	0.99695	0.00109	0.96943	0.01175	0.09612
4.0	0.3597	0.99792	0.0^3656	0.97618	0.00809	0.07564
5	0.2878	0.90890	0.0^3279	0.98438	0.00430	0.05028
6	0.2398	0.99935	0.0^3138	0.98898	0.00255	0.03580
7	0.2055	0.99959	0.0^4758	0.99181	0.00164	0.02677
8	0.1799	0.99972	0.0^4450	0.99368	0.00111	0.02077
9	0.1599	0.99980	0.0^4284	0.99496	0.0^3788	0.01658
10	0.1439	0.99985	0.0^4188	0.99590	0.0^3579	0.01354
15	0.0959	0.9^455	0.0^5380	0.99815	0.0^3176	0.00617
20	0.0719	0.9^480	0.0^5122	0.99895	0.0^4751	0.00351
30	0.0480	0.9^543	0.0^6244	0.99953	0.0^4225	0.00158
40	0.0360	0.9^575	0.0^7776	0.99974	0.0^5956	0.0^3894
50	0.0288	0.9^588	0.0^7319	0.99983	0.0^5491	0.0^3574
100	0.0144	0.9^685	0.0^8201	0.99996	0.0^6619	0.0^3144
	small x	$1 - 0.0513x^3$	$0.0472x^4$	$1 - 0.208x^2$	$0.2092x^3$	$0.7035x^2$

Source: Adapted, with permission, from C. W. Allen, *Astrophysical Quantities*, Athlone Press, London, 1973, pp. 104–106.

[a] The radiance L (also known as the specific intensity or brightness) is defined as the flux per unit projected area per unit solid angle leaving a surface. The total flux leaving a surface is denoted by M and is called the exitance (or the emittance). For blackbody radiation, L and M are related by $M = \pi L$. Absolute numerical values of the quantities appearing in this table can be obtained through use of the relations in Table 4.2.

Table 4.2. Absolute Numerical Values of the Quantities Appearing in Table 4.1

$$x = \frac{h\nu}{kT} = \frac{hc}{k\lambda T} = \frac{1.4388 \text{ cm K}}{\lambda T} = \frac{0.014388 \text{ m K}}{\lambda T}$$

$$\int_0^\infty L_\lambda \, d\lambda = 1.8047 \times 10^{-5} T^4 \text{ erg/cm}^2 \text{ s sr K}^4$$
$$= 1.8047 \times 10^{-8} T^4 \text{ W/m}^2 \text{ sr K}^4$$

$$L_{\lambda, \max} = 4.0951 \times 10^{-5} T^5 \text{ erg/cm}^3 \text{ s sr K}^5$$
$$= 4.0951 \times 10^{-6} T^5 \text{ W/m}^3 \text{ sr K}^5$$

$$\int_0^\infty L_{p, \lambda} \, d\lambda = 4.8396 \times 10^{10} T^3 \text{ photons/cm}^2 \text{ s sr K}^3$$
$$= 4.8396 \times 10^{14} T^3 \text{ photons/m}^2 \text{ s sr K}^3$$

$$L_{p, \lambda, \max} = 6.6871 \times 10^{10} T^4 \text{ photons/cm}^3 \text{ s sr K}^4$$
$$= 6.6781 \times 10^{16} T^4 \text{ photons/m}^3 \text{ s sr K}^4$$

$$L_{\nu, \max} = 1.8959 \times 10^{-16} T^3 \text{ erg/cm}^2 \text{ sr K}^3$$
$$= 1.8959 \times 10^{-19} T^3 \text{ W/m}^2 \text{ Hz sr K}^3$$

which has the numerical value

$$L_{\lambda, \max}(T) = 4.095 \times 10^{-6} T^5 \left(\frac{1}{K^5} \right) \left(\frac{W}{m^2 \text{ m sr}} \right), \qquad (4.21)$$

and to identify the normalized spectral distribution function as

$$f_\lambda \left(\frac{\lambda kT}{hc} \right) = \frac{1}{21.20} \frac{(hc/\lambda kT)^5}{e^{hc/\lambda kT} - 1}. \qquad (4.22)$$

This function is plotted in Fig. 4.2 both as a function of $\lambda kT/hc$ and as a function of λT in units of m K.

In many circumstances, it is desirable to evaluate the Planck radiation functions through the use of tabulated values. Such values can be obtained from Tables 4.1 and 4.2 which also summarizes the results of this section.

4.2 RADIATION FROM NONBLACKBODY SOURCES

The spectral radiance of a blackbody emitter of temperature T can be calculated from the Planck radiation law [Eq. (4.1) or Eq. (4.12)]. An actual source of temperature T produces a radiance that, in general, is smaller than that predicted by the Planck law. The ratio of the actual radiance to the predicted radiance is known as the emissivity of the source. [The emissivity will be defined more precisely later in this section in connection with Eqs. (4.30)–(4.32)].

An actual source that is designed to mimic the properties of an ideal blackbody is known as a simulated blackbody. The emissivity of a simulated blackbody can readily exceed 0.95. Details regarding the construction of a simulated blackbody can be found in Section 3.1 of the book by Hudson included in the references at the end of this chapter.

Phenomenological Description of Radiation Sources

Consider the object shown in Fig. 4.3. A beam containing an amount of flux Φ_0 is incident on the material, of which an amount Φ_1 is reflected, Φ_2 is absorbed, and Φ_3 is transmitted. The reflectance R of the object is defined as the fractional transmitted flux

$$R = \frac{\Phi_1}{\Phi_0}, \tag{4.23}$$

the absorptance α is defined as the fractional absorbed flux

$$\alpha = \frac{\Phi_2}{\Phi_0}, \tag{4.24}$$

and the transmittance T is defined as the fractional transmitted flux

$$T = \frac{\Phi_3}{\Phi_0}. \tag{4.25}$$

Energy conservation requires that all of the incident flux be accounted for, or that

$$R + T + \alpha = 1. \tag{4.26}$$

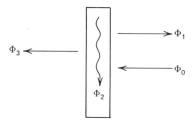

Figure 4.3. Incident, reflected, absorbed, and transmitted fluxes.

If Kirchhoff's law $\varepsilon = \alpha$ can be assumed valid for each spectral and directional component of the incident radiation [see the discussion following Eq. (3.16)], the emissivity ε of the object for the direction from which the radiation is incident, and for the same spectral interval as that comprising the incident radiation, must be given by

$$\varepsilon = 1 - R - T. \tag{4.27}$$

(For the special case of an opaque object, i.e., one sufficiently thick that $T = 0$, the quantities R and α are sometimes referred to as the reflectivity and absorptivity, respectively. The emissivity is then given, assuming the validity of Kirchhoff's law, by $\varepsilon = 1 - R$. Different nomenclature is used in this case to indicate that α and R are intrinsic properties of the material and not of its geometry.)

Radiation Temperature

Figure 4.4 shows a plot of the spectral radiance of blackbodies of several different temperatures. An important feature of the Planck radiation law which is illustrated by this figure is that curves for different temperatures do not cross. Thus the value of the Planck function at a single wavelength uniquely defines the temperature of the source. Even a source that is not a blackbody can be described at any wavelength λ by its radiation temperature T_R, defined as the temperature for which the Planck radiation law evaluated at wavelength λ is equal to the source radiance at wavelength λ. This requires that

$$L_\lambda(\lambda) = \frac{2hc^2}{\lambda^5} \frac{1}{e^{hc/\lambda k T_R} - 1}, \tag{4.28}$$

which can be solved for T_R:

$$T_R = \frac{hc}{\lambda k \ln\left[1 + 2hc^2/\lambda^5 L_\lambda(\lambda)\right]}. \tag{4.29}$$

Emissivity

Thus far the word emissivity has been used in a general sense to describe the ratio of the flux emitted by an object to that emitted by a blackbody emitter of the same geometry. By specifying exactly the conditions under which the flux is to be measured, a more precise definition of the emissivity can be given.

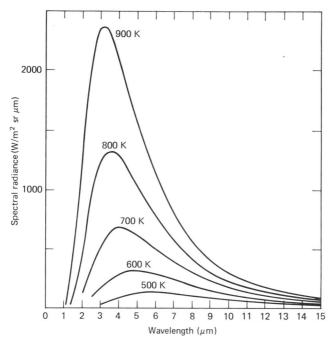

Figure 4.4. Planck radiation law for several blackbody temperatures. (Used with permission from R. D. Hudson, *Infrared System Engineering*, Wiley, New York, 1969.)

The ratio of the radiance of a thermal source to that of a blackbody of the same temperature defines the spectral goniometric emissivity

$$\varepsilon(\theta, \phi, \lambda, T) = \frac{L_\lambda(\theta, \phi, \lambda, T)|_{\text{source}}}{L_\lambda(\theta, \phi, \lambda, T)|_{\text{blackbody}}}. \qquad (4.30)$$

Here θ and ϕ are angles that define the direction in which the radiance is to be measured. Under certain circumstances, a knowledge of the directional properties of the emitted radiation may be unimportant, and in these cases the source can be described by its spectral hemispherical emissivity, defined as the ratio of the spectral exitance of the source to that of a blackbody of the same temperature:

$$\varepsilon(\lambda, T) = \frac{M_\lambda(\lambda, T)|_{\text{source}}}{M_\lambda(\lambda, T)|_{\text{blackbody}}}. \qquad (4.31)$$

Finally, if knowledge of both spectral and directional properties of the radiation are unimportant, the source can be described by its hemispherical emissivity, which is defined as the ratio of the source exitance to that of a blackbody of the same temperature:

$$\varepsilon(T) = \frac{M(T)|_{\text{source}}}{M(T)|_{\text{blackbody}}} = \frac{M(T)|_{\text{source}}}{\sigma T^4}. \tag{4.32}$$

This quantity corresponds most closely to the quantity introduced in Section 3.2.

4.3 RADIATIVE HEAT TRANSFER BETWEEN SURFACES

In many practical applications, it is necessary to calculate the rate at which heat is transferred between objects by thermal radiation. An example of such a calculation is presented in this section. Other examples can be found in the problems at the end of this chapter.

Figure 4.5 shows two plane parallel plates having emissivities ε_1 and ε_2 and temperatures T_1 and T_2. Assuming that the distance between the plates is small compared to their linear dimensions, we can calculate the rate per unit area at which heat is transferred between the plates. It is assumed that the region between the plates is evacuated or contains a lossless dielectric.

Problems of this sort are solved most readily by considering the total radiation field in the region between the plates to be composed of two components, one of which is directed toward surface 1, the other being directed toward surface 2. The first component produces an irradiance E_1 on surface 1 while the second component produces an irradiance E_2 on surface 2. The irradiance of surface 2 can be considered to result from two contributions: The first has magnitude $\varepsilon_1 \sigma T_1^4$ and results from direct irradiation by surface 1, and the second has magnitude $E_1(1 - \varepsilon_1)$ and results from the partial reflection of the flux incident on surface 1. The irradiance of surface 2 can thus be expressed as

$$E_2 = \varepsilon_1 \sigma T_1^4 + E_1(1 - \varepsilon_1), \tag{4.33}$$

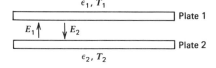

Figure 4.5. Two plane parallel plates transfer heat by radiation.

and similarly the irradiance of surface 1 can be expressed as

$$E_1 = \varepsilon_2 \sigma T_2^4 + E_2(1 - \varepsilon_2).$$ (4.34)

The net rate per unit area at which heat is transferred from surface 1 to surface 2 must be given by the difference between these expressions, or as

$$E_{21} = E_2 - E_1.$$ (4.35)

Equations (4.33)–(4.35) can be solved to give the result:

$$E_{21} = \frac{\sigma\left(T_1^4 - T_2^4\right)}{1/\varepsilon_1 + 1/\varepsilon_2 - 1}.$$ (4.36)

It can be noted that the rate of heat transfer is maximum if both surfaces are black (i.e., if $\varepsilon_1 = \varepsilon_2 = 1$) and that this rate decreases as the emissivity of either surface decreases. An alternative method of calculating the rate of heat flow would be to follow individual rays of radiation as they undergo repeated reflections between the two surfaces. Treated in this manner, the net calculation of the net rate would require a summation of the energy transferred by each of the infinity of reflections. The procedure followed here automatically accounts for this infinity of reflections.

4.4 RADIATIVE TRANSFER THROUGH ABSORBING MEDIA

The previous discussion has dealt with radiative transfer through a vacuum or through other nonabsorbing media. The theory of radiative transfer through absorbing media is much more complicated. In this case, the radiation is continually being absorbed and reemitted, and thus the radiance of this radiation can change continuously with position and with direction. It should be noted that the radiance theorem is not valid for transfer through absorbing media. Historically, considerations of radiative transfer through absorbing media were motivated by attempts to understand the properties of stellar atmospheres.

The equation of radiative transfer can be derived from a consideration of the cylindrical region of base area dA and of height dz shown in Fig. 4.6. Radiation contained in the solid angle $d\Omega$ falls at normal incidence onto this cylinder. In passing through the cylinder, the radiance of the incident beam decreases through absorption by an amount $-\kappa_\nu L_\nu\, dz$, where κ_ν is the absorption coefficient for radiation of frequency ν [cf. Eq. (1.31)]. The radiance of the beam can also increase, however, by emission from the material contained within the cylinder. Let us assume that $j_\nu/4\pi$ denotes the rate per unit volume per unit solid angle per unit frequency interval at which energy is emitted by this material. The radiance is then increased in passing through the cylinder by an amount $j_\nu\,dz/4\pi$. The net change in radiance is the sum of these contributions, or

$$dL_\nu = -\kappa_\nu L_\nu\,dz + \frac{j\nu}{4\pi}\,dz. \tag{4.37}$$

This equation can be reexpressed by defining the optical depth τ_ν such that

$$d\tau_\nu = \kappa_\nu\,dz \tag{4.38}$$

and defining a source function $\mathcal{J}_\nu$ by

$$\mathcal{J}_\nu = \frac{j_\nu}{4\pi\kappa_\nu}. \tag{4.39}$$

Substitution of these quantities yields the transfer equation

$$\frac{dL_\nu}{d\tau_\nu} = -L_\nu + \mathcal{J}_\nu. \tag{4.40}$$

The quantities L_ν and $\mathcal{J}_\nu$ depend, in general, on the direction of propagation of the radiation, but this dependence is suppressed in our notation.

An important special case of Eq. (4.40) can be obtained by assuming that j_ν and κ_ν are related in the same way that they would be related if the

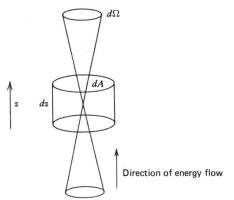

Direction of energy flow

Figure 4.6. Geometry used in deriving the equation of radiative transfer.

material and the radiation were strictly in thermodynamic equilibrium. This assumption is known as the assumption of *local thermodynamic equilibrium* (LTE), and plays an important role in the theory of stellar atmospheres (see, e.g., the discussion in Chandrasekhar or Mihalis). The form of the relation between j_ν and κ_ν can be derived by considering the energy balance of the absorption and emission processes in thermodynamic equilibrium. The total flux absorbed by the volume element of Fig. 4.6 from radiation within the solid angle $d\Omega$ is given by

$$\kappa_\nu L_\nu^{\mathrm{BB}}(T)\, dA\, dz\, d\Omega, \tag{4.41}$$

where $L_\nu^{\mathrm{BB}}(T)$ denotes the spectral radiance of a blackbody of temperature T. The total flux emitted into this solid angle is given by

$$j_\nu\, dA\, dz\, \frac{d\Omega}{4\pi}. \tag{4.42}$$

Equating these expressions yields the result

$$j_\nu = 4\pi\kappa_\nu L_\nu^{\mathrm{BB}}(T). \tag{4.43}$$

This equation constitutes an alternative statement of Kirchhoff's law [cf. Eq. (3.16)]. A consequence of Eqs. (4.39) and (4.43) is that the source function can be expressed as

$$\mathcal{J}_\nu(\tau_\nu) = L_\nu^{\mathrm{BB}}[T(\tau_\nu)] \tag{4.44}$$

where $T(\tau_\nu)$ denotes the temperature at an optical depth of τ_ν.

The solution of the equation of radiative transfer (4.40) is extremely difficult under most conditions. Some solutions of this equation are discussed by Chandrasekhar and by Sobolev in the references listed at the end of this chapter.

A simple example of the application of the transfer equation is illustrated in Fig. 4.7. Radiation is emitted by a uniform slab of material assumed to be in local thermodynamic equilibrium at the temperature T. The radiance of the outgoing radiation in a direction normal to faces of the slab is to be calculated. Under these circumstances, the transfer equation (4.40) becomes, using the source term of Eq. (4.44),

$$\frac{dL_\nu(\tau_\nu)}{d\tau_\nu} = -L_\nu(\tau_\nu) + L_\nu^{\mathrm{BB}}(T). \tag{4.45}$$

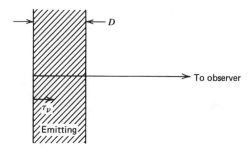

Figure 4.7. Calculation of the radiance of a uniform slab in local thermodynamic equilibrium.

A boundary condition of the form

$$L_\nu(0) = 0 \tag{4.46}$$

is assumed, implying that no radiation is incident on the slab. The solution to Eq. (4.45) under this condition is given by

$$L_\nu(\tau_\nu) = L_\nu^{BB}(T)(1 - e^{-\tau_\nu}). \tag{4.47}$$

It can be noted that in the limit of an optically thick source (i.e., for $\tau_\nu \gg 1$) the radiance approaches that of a blackbody. In the opposite limit of an optically thin source (i.e., for $\tau_\nu \ll 1$) the radiance depends linearly on the optical depth.

BIBLIOGRAPHY

Thermal Sources

C. W. Allen, *Astrophysical Quantities*, Athlone Press, London, 1973, pp. 104–106.

R. D. Hudson, Jr., *Infrared System Engineering*, Wiley, New York, 1969, Chapters 2 and 3.

P. W. Kruse, L. D. McGlauchlin, and R. B. McQuistan, *Infrared Technology*, Wiley, New York, 1962, Chapters 2 and 3.

W. L. Wolfe and G. J. Zissis, *The Infrared Handbook*, ONR Washington, D.C., 1978, Chapter 1.

Radiative Transfer

L. H. Aller, *Astrophysics*, Ronald, New York, 1953, Chapters 5 and 7.

S. Chandrasekhar, *Radiative Transfer*, Oxford University, London, 1950.

H. C. Hottel and A. F. Sarofim, *Radiative Transfer*, McGraw-Hill, New York, 1967.

D. Mihalis *Stellar Atmospheres*, Freeman, San Francisco, 1970.

V. V. Sobolev, *A Treatise on Radiative Transfer*, Van Nostrand, Princeton, N.J., 1963.

E. M. Sparrow and R. D. Cess, *Radiation Heat Transfer*, Brooks-Cole, Belmont, Calif., 1966.

J. A. Wiebelt, *Engineering Radiation Heat Transfer*, Holt, Rinehart and Winston, New York, 1966.

PROBLEMS

1 It was noted in the text that the maximum values of L_ν and of L_λ occur at different values of ν (or of $\lambda = c/\nu$).

   At what frequency does the radiance per unit logarithmic frequency interval, which can be denoted $L_{\ln \nu}$, have its maximum value? Similarly, at what frequency does $L_{\ln \lambda}$ have its maximum value? What is the physical significance of $L_{\ln \lambda}$?

2 For a given value of the temperature T, find the frequency ν_{max} at which the photon radiance per unit frequency interval, $L_{p,\nu}$, is a maximum. (Photon radiance is defined to be the rate per unit projected area at which photons are emitted into unit solid angle). Show that it is possible to express $L_{p,\nu}$ in the form $L_{p,\nu} = L_{p,\nu,max}(T)f(h\nu/kT)$. Find explicit forms for $L_{p,\nu,max}(T)$ and $f(h\nu/kT)$. Numerically evaluate $L_{p,\nu,max}$ and sketch $f(h\nu/kT)$.

3 A sphere of radius R_0 is suspended by wires of negligible thermal conductivity inside a large evacuated enclosure whose walls have an emissivity of unity and are maintained at temperature T. The sphere has an emissivity of ε_0 and is heated electrically with a power P_0. What is its equilibrium temperature T_0?

   A radiation shield consisting of a thin, spherical shell of metal, concentric with the sphere, is now inserted between the sphere and the enclosure. The shell has radius R_1 and emissivity ε_1. That fraction of the radiation incident on the shell that is not absorbed is specularly reflected. Assume that for both the sphere and the shell the thermally emitted radiation obeys Lambert's law. What is the equilibrium temperature T_1 of the shell and what is the new temperature of the sphere?

   Evaluate your results for the case of $R_0 = 1$ cm, $R_1 = 2$ cm, $\varepsilon_0 = 0.5$, $\varepsilon_1 = 0.1$, $P_0 = 1$ W, and $T = 300$ K.

4 How would the results of Problem 3 differ if it had been assumed that the fraction of the radiation incident on the inner shell that is not absorbed is diffusely reflected in accordance with Lambert's law?

5 Consider a spherical shell of luminous gas in local thermodynamic equilibrium at a constant temperature T. As shown in the following

figure, the outer diameter is R and the shell thickness is ΔR. The absorption coefficient is denoted κ. Consider the distribution of radiance across the surface of the shell, that is, determine the appearance of this shell to a distant observer.

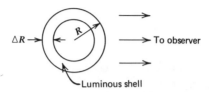

Sketch the radiance distribution for the following cases:

(a) $\Delta R/R = 0.1$, $\kappa \Delta R = 0.01$;

(b) $\Delta R/R = 0.1$, $\kappa \Delta R = 10$;

(c) $\Delta R/R = 1$, $\kappa \Delta R = 0.01$;

(d) $\Delta R/R = 1$, $\kappa \Delta R = 10$.

5

Advanced Topics
in Radiometry

In this chapter, the formal development of radiometry begun in Chapter 2 is continued. The laws of radiometry are derived here by two essentially distinct methods. The more straightforward method treats the laws of radiometry as a consequence of the laws of geometrical optics. The other method considers the propagation of blackbody radiation through optical systems and thus deduces the laws of radiometry from the laws of thermodynamics. Both approaches are used in this chapter, since each provides a useful physical picture of the nature of these laws. Also included in this chapter are a brief review of geometrical optics and discussions of the radiometric properties of image-forming systems.

5.1 GEOMETRICAL OPTICS AND IMAGE FORMATION

The present section presents a brief review of those aspects of geometrical optics that are necessary to an understanding of the radiometric properties of image-forming optical systems. Most of the results presented here will be stated without proof; readers requiring additional information can consult the references listed at the end of this chapter.

Geometrical Optics

The laws of geometrical optics are based on the application of the laws regarding the reflection and refraction of light, which can be stated with reference to the diagrams shown in Fig. 5.1. The law of reflection states that a ray of light reflects from an interface in such a way that the reflected ray lies in the place of incidence with the angle of reflection ϕ equal to the angle of incidence θ. Here the plane of incidence is defined to be the plane containing the surface normal $\hat{N}$ and the incident ray. This plane is

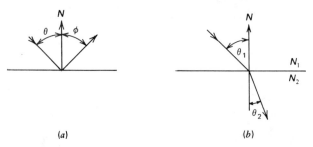

(a)   (b)

Figure 5.1. (a) Law of reflection. (b) Law of refraction, (Snell's law).

conveniently taken to be the plane of the paper of Fig. 5.1. The law of refraction, also known as Snell's law, states that, on encountering a boundary between regions of refractive indices N_1 and N_2, a ray of light is refracted such that it lies in the plane of incidence with the angle of refraction θ_2 related to the angle of incidence θ_1 by

$$N_1 \sin \theta_1 = N_2 \sin \theta_2. \tag{5.1}$$

A simple example of an optical imaging system is a thin lens located in air, as shown in Fig. 5.2. The lens can be considered thin as long as its thickness d is much less than the radii of curvature of the lens surfaces, denoted r_1 and r_2. A thin lens can be characterized by its focal length f, given by the lens maker's formula as

$$\frac{1}{f} = (N - 1)\left(\frac{1}{r_1} + \frac{1}{r_2}\right). \tag{5.2}$$

The use of a thin lens to form an image is illustrated in Fig. 5.3. The object distance s_1 is related to the image distance s_2 by the lens formula

$$\frac{1}{f} = \frac{1}{s_1} + \frac{1}{s_2}. \tag{5.3}$$

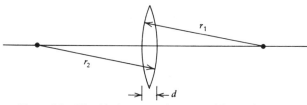

Figure 5.2. The thin lens; r_1 and r_2 are positive as shown.

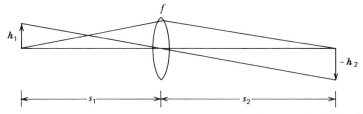

Figure 5.3.   A thin lens of focal length f forms an image of height $-h_2$ of an object of height h_1.

Near the axis, a lens is approximately a plane parallel plate, and thus an undeviated ray connects the tip of the object to the tip of the image. The magnification of this system, defined as $m \equiv h_2/h_1$, is thus given by $m = -s_2/s_1$.

Thick Lenses

A general optical system may be regarded as a thick lens. Thus a thick lens can be either a single lens of arbitrary thickness or a combination containing many individual elements. An example, in the form of a thick lens composed of a single element, is shown in Fig. 5.4. Points F_1 and F_2 are known as the first and second focal points, respectively. Light leaving F_1 and traveling to the right is rendered parallel by the lens, while the parallel light incident on the lens from the left is brought to a focus at F_2. The points H_1 and H_2 are known as the first and second principal points, and planes perpendicular to the optical axis through these points are known as principal planes. The significance of these planes is that, as shown in the figure, extensions of the segments of the ray leaving F_1 intersect on the first principal plane, while extensions of the segments of the ray directed at F_2 intersect on the second principal plane. The distance from F_1 to H_1 is known as the first focal length and is designated f_1, while the distance from H_2 to F_2 is known as the second focal length and is designated f_2. In general, these focal lengths are related by

$$\frac{f_1}{N_1} = \frac{f_2}{N_2}. \tag{5.4}$$

The discussion thus far in this paragraph applies to any optical system. For the special case of the single-element thick lens shown in Fig. 5.4, these focal

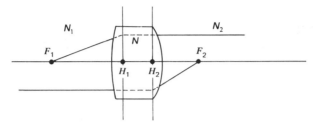

Figure 5.4. A thick lens.

lengths are given by

$$\frac{f_1}{N_1} = \frac{f_2}{N_2} = \frac{N - N_1}{r_1} + \frac{N - N_2}{r_2} - \frac{d(N - N_1)(N - N_2)}{Nr_1r_2}, \quad (5.5)$$

where the quantities r_1 and r_2 are the radii of curvature of the front and back surfaces and are positive values for the lens shown.

The general imaging properties of any thick lens are illustrated in Fig. 5.5. The lens has been omitted entirely from this figure, since its properties are completely specified by the positions of its focal points and principal points. An object at P_1 is imaged to the point P_2. If the object and image distances s_1 and s_2 are measured to their respective principal planes, the imaging law takes the form

$$\frac{N_1}{s_1} + \frac{N_2}{s_2} = \frac{N_1}{f_1} = \frac{N_2}{f_2}. \quad (5.6)$$

Stops and Apertures

The presence of limiting apertures, or stops, plays a key role in determining the radiometric properties of imaging systems, since they limit the

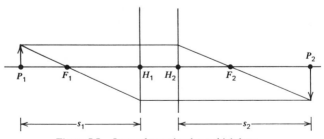

Figure 5.5. Image formation by a thick lens.

amount of flux that can be transmitted by the system. A stop can be either an aperture purposely introduced into the system for this purpose or the clear aperture of some lens.

The *aperture stop* of an optical imaging system is the boundary that limits the cone angle of rays leaving an object point on axis and being transmitted by the system. In the example shown in Fig. 5.6, the aperture stop takes the form of a physical stop placed in contact with the first lens. The image of the aperture stop in object space is known as the *entrance pupil*, while the image in image space is known as the *exit pupil*. Since all of the light transmitted by the system must pass through the aperture stop, all of the light in the image space appears to pass through the exit pupil. Thus in an optical system intended for visual use, the observer's eye ideally should be placed at the exit pupil.

The field of view of an imaging system is limited by an aperture known as the *field stop*. The field stop may be determined by the following method: Any ray that passes through the center of the aperture stop is known as a *chief ray*. This ray will also pass through the centers of the entrance and exit pupils. Different chief rays thus correspond to different object and image points. Some aperture of the system will limit the maximum cone angle of chief rays that are transmitted by the system, and this aperture is known as the field stop. The *entrance window* is the image of the field stop in object space, while the *exit window* is the image in image space. The angle subtended by the exit window at the exit pupil gives the field of view of the system. If the field stop coincides with an image plane of the optical system, the entrance window will coincide with the object plane. If this is not the case, different portions of the object field will be visible from different points within the exit pupil, due to parallax. This effect leads to vignetting, a decrease in image irradiance at off-axis image points.

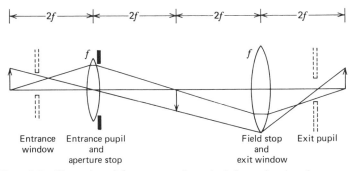

Figure 5.6. Illustration of the stops, pupils, and windows of an imaging system.

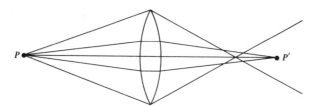

Figure 5.7. Spherical aberration.

Aberrations

The preceding discussion of image formation has implicitly assumed what is kown as the paraxial approximation (i.e., all of the rays under consideration lie close to the optical axis of the optical system). More explicitly, this approximation requires that the angle between any ray and the optical axis be small compared to unity and that the height of any ray as it intersects an optical element be small compared to the focal length of that element. Under these conditions, imaging is perfect in that all of the rays passing through the system from a fixed object point will converge on a single point in image space.

Outside the paraxial approximation, imaging need not be perfect. The biconvex thin lens shown in Fig. 5.7, for example, suffers from spherical aberration in that not all of the rays leaving the point P converge on P'. In the simple example shown, where the lens is used at one-to-one imaging, the radius of the blur circle at the paraxial image plane can be shown to be given approximately (i.e., correct to third order in D/f) by

$$\Delta y = \frac{D^3}{32 f^2} \frac{N^2}{(N-1)^2}, \tag{5.7}$$

where f and D represent the focal length and diameter of the lens whose refractive index is taken as N.

An image-forming system, such as that shown in Fig. 5.8, is said to be *stigmatic* for the two axial field points P and P' if P' is a perfect image of P,

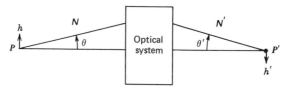

Figure 5.8. Definition of distances and angles related by Abbe's sine condition.

that is, if the system does not have spherical aberration. This condition does not ensure, however, that points slightly off axis will form perfect images. For instance, the system can display an aberration known as coma, which results when different annular regions of the lens produce different linear magnifications. In order for a system that is stigmatic for points P and P' to be free of coma as well, it is necessary that the system obey the sine condition of Abbe [E. Abbe, *Jena. Z. Med. Naturwiss.*, 129 (1879); *Carl. Repert. Phys.*, **16**, 303 (1880)] which requires that

$$Nh \sin \theta = N'h' \sin \theta' \qquad (5.8)$$

for all axial rays. Here N and N' are the refractive indices of the object and image spaces, h and h' are the object and image heights, and θ and θ' are the angles of an axial ray in object and image spaces, respectively. It is assumed here that h and h' are small, but that θ and θ' may be arbitrarily large. A system that is stigmatic for two points P and P' and also obeys the sine condition is said to be *aplanatic* for the two points. A remarkable feature of Abbe's sine condition is that it relates the quality of the image of an off-axis point to the properties of axial rays.

Other aberrations that can be present in imaging systems include chromatic aberration, astigmatism, distortion, and field curvature. These will not be discussed here since they do not enter the following development of the theory of radiometry.

5.2 RADIANCE THEOREM

It was shown in Chapter 2 that the radiance L of any narrow beam of light is conserved as the light propagates through a uniform, lossless medium. We shall now demonstrate the general form of this result: The *basic radiance* (defined as L/N^2, where N is the refractive index of the medium) of a narrow beam of radiation is conserved as the beam propagates through any lossless optical system. We shall first show that the quantity L/N^2 is conserved as the beam propagates across boundaries separating uniform media; the general result then follows from recognizing that an arbitrarily complicated optical system is composed of a sequence of uniform media separated by boundaries.

Geometrical Proof of Radiance Theorem

Figure 5.9 shows an area element dA on the surface S separating optical media of refractive indices N_1 and N_2. A beam of radiance L_1 falls onto dA

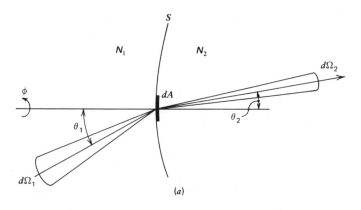

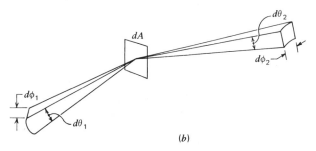

Figure 5.9. Radiance theorem.

from the element $d\Omega_1$ of solid angle, inclined at the angle θ_1 to the normal to dA. The flux carried by this beam is thus given by

$$d^2\Phi = L_1\, dA \cos\theta_1\, d\Omega_1. \tag{5.9}$$

In order to determine the radiance of the beam leaving the surface, expressions for θ_2 and $d\Omega_2$ in terms of θ_1 and $d\Omega_1$ are needed. In a polar coordinate system whose axis is the normal to dA, the ratio of the solid angle elements $d\Omega_1$ and $d\Omega_2$ is given by

$$\frac{d\Omega_1}{d\Omega_2} = \frac{\sin\theta_1\, d\theta_1\, d\phi_1}{\sin\theta_2\, d\theta_2\, d\phi_2}, \tag{5.10}$$

where $d\phi_1$ and $d\phi_2$ are elements of the aximuthal angle, as shown in Fig. 5.1b. Snell's law states that a refracted ray must remain in the plane of incidence, so that

$$d\phi_1 = d\phi_2,$$ (5.11)

and the angles of incidence and refraction are related by

$$N_1 \sin \theta_1 = N_2 \sin \theta_2,$$ (5.12)

which can be differentiated to obtain

$$N_1 \cos \theta_1 \, d\theta_1 = N_2 \cos \theta_2 \, d\theta_2.$$ (5.13)

Equation (5.10) can be simplified through use of these results to give

$$\frac{d\Omega_1}{d\Omega_2} = \frac{N_2^2 \cos \theta_2}{N_1^2 \cos \theta_1}.$$ (5.14)

Finally, the radiance of the refracted beam is given by

$$L_2 = \frac{d^2 \Phi}{dA \cos \theta_2 \, d\Omega_2}$$

$$= L_1 \frac{dA \cos \theta_1 \, d\Omega_1}{dA \cos \theta_2 \, d\Omega_2}$$

$$= L_1 \frac{N_2^2}{N_1^2},$$ (5.15)

and thus

$$\frac{L_1}{N_1^2} = \frac{L_2}{N_2^2}.$$ (5.16)

Since it was shown previously that the radiance is conserved for propagation through a uniform medium, it can be concluded that as a ray propagates through an optical system the basic radiance at each point along the ray measured in the direction of the ray is an invariant quantity.

Radiance Theorem and Blackbody Radiation

It is interesting that the radiance theorem can be deduced from the properties of blackbody radiation. Figure 5.10 shows two isothermal en-

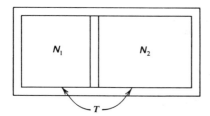

Figure 5.10. Radiance theorem and blackbody radiation.

closures held at the common temperature T. The enclosures are filled with dielectrics of refractive indices N_1 and N_2, respectively. The basic spectral radiances of the blackbody radiation filling each cavity are given from Eq. (3.69) as

$$\frac{L_{\nu 1}}{N_1^2} = \frac{L_{\nu 2}}{N_2^2} = \frac{2h\nu^3}{c^2} \frac{1}{e^{h\nu/kT} - 1}. \qquad (5.17)$$

Suppose that a small hole is now drilled into the wall separating the two cavities so that they can interchange radiation. This process cannot disturb the radiation field in either cavity, since they are still in thermal equilibrium. The radiance of the radiation must therefore change from $L_{\nu 1}$ to $L_{\nu 2} = L_{\nu 1} N_2^2 / N_1^2$ in passing to the right through the interface and must change from $L_{\nu 2}$ to $L_{\nu 1}$ in passing in the opposite direction. Thus the basic spectral radiance L_{ν}/N^2 is conserved as radiation passes through a dielectric interface.

This demonstration relies on the fact that Planck's radiation law predicts that the spectral radiance of blackbody radiation will be proportional to the square of the refractive index of the medium that contains the radiation field. In fact, this conclusion can be deduced solely from the laws of thermodynamics, as was demonstrated by Kirchhoff and independently by Clausius (see, e.g., pp. 502–505 in the account in Drude, mentioned in the bibliography at the end of this chapter). Figure 5.11 shows two infinite plane surfaces P and P' that are blackbody emitters of temperature T. Plane P is in contact with a dielectric medium of refractive index N and P' is in contact with a medium of index $N' > N$. The surfaces P and P' and the interface between the dielectrics are all parallel. We denote by L the radiance of P as measured in N and by L' the radiance of P' as measured in N'. Each area element dA on P radiates an element of flux

$$d^2\Phi = 2\pi L \, dA \cos\theta \sin\theta \, d\theta \qquad (5.18)$$

into an annular element of solid angle. A fraction $r(\theta)$ of this flux is

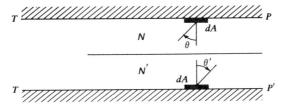

Figure 5.11. Thermodynamic demonstration of the radiance theorem.

reflected at the interface and is reabsorbed by P, and a fraction $1 - r(\theta)$ is transmitted by the interface and is absorbed by P'. Similarly, each area element dA on P' radiates a flux

$$d^2\Phi = 2\pi L' \, dA \cos\theta' \sin\theta' \, d\theta' \tag{5.19}$$

into an annular element of solid angle, a fraction $1 - r'(\theta')$ of which is absorbed by P. For equilibrium to hold, each element dA of P must transfer as much flux to P' as each element of P' transfers to P, requiring that

$$\int_0^{\pi/2} L[1 - r(\theta)]\sin\theta \cos\theta \, d\theta$$

$$= \int_0^{\arcsin(N/N')} L'[1 - r'(\theta')]\sin\theta' \cos\theta' \, d'\theta'. \tag{5.20}$$

The upper limit of integration of the right-hand side of this expression runs only to $\arcsin(N/N')$ since, for $\sin\theta' > N/N'$, total internal reflection occurs and $r'(\theta') = 1$. We now make the substitution $\sin\theta' = (N/N')\sin\theta$ in this integral, which has the effect of expressing the θ' integration in terms of the angle θ which is related to θ' through Snell's law. Microscopic reversibility then requires that $r(\theta) = r'(\theta')$, and thus Eq. (5.20) becomes

$$L\int_0^{\pi/2}[1 - r(\theta)]\sin\theta \cos\theta \, d\theta$$

$$= \frac{N^2}{N'^2} L' \int_0^{\pi/2}[1 - r(\theta)]\sin\theta \cos\theta \, d\theta. \tag{5.21}$$

Since the integral does not vanish, we can conclude that the radiances L and L' measured in media of indices N and N', respectively, are related through

$$\frac{L}{N^2} = \frac{L'}{N'^2}. \tag{5.22}$$

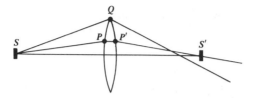

Figure 5.12.   Optical system showing spherical aberration.

5.3   THE RADIANCE OF IMAGES

Let us now consider the problem of calculating the radiance of the image of a source. In one sense, the solution of this problem follows trivially from the radiance theorem, which states that the basic radiance is conserved along any ray. A subtlety is introduced, however, by the possible presence of aberrations. In Fig. 5.12, S' is the paraxial image of a small, uniform Lambertian source S. Thus the radiance of the image in the direction $P'S'$ of a paraxial ray is equal to that of the source. Spherical aberration can cause the marginal ray SQ to miss the paraxial image, however, and thus the radiance of S' in the direction QS' is equal to zero. If an observer were to place his eye at S' and look back at the lens, only the central portion of the lens would appear illuminated by S.

Hence a discussion of the radiance of an image must be based on a knowledge of the image quality. Therefore, let us consider the optical system shown in Fig. 5.13, which is assumed to obey the sine condition. We wish to calculate the radiance of the image of a source element $dA_0 = dx_0\, dy_0$. The flux radiated by dA_0 into an element of solid angle $d\Omega_0 = \sin\theta_0\, d\theta_0\, d\phi$ at the entrance pupil of the optical system is given by

$$d^2\Phi = L_0(\theta_0, \phi)\, dx_0\, dy_0 \sin\theta_0 \cos\theta_0\, d\theta_0\, d\phi. \qquad (5.23)$$

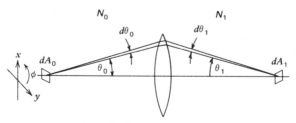

Figure 5.13.   Radiance of the image of dA_0.

If the system is lossless, this flux must pass through an area element $dA_1 = dx_1\,dy_1$ in the image space from a solid angle $d\Omega_1 = \sin\theta_1\,d\theta_1\,d\phi$, and thus the image radiance in this direction is given by

$$L_1(\theta_1, \phi) = \frac{d^2\Phi}{dA_1 \cos\theta_1\,d\Omega_1}. \tag{5.24}$$

By the sine condition, we have

$$N_0\,dx_0 \sin\theta_0 = N_1\,dx_1 \sin\theta_1, \tag{5.25}$$

and by changing x to y and differentiating, we have

$$N_0\,dy_0 \cos\theta_0\,d\theta_0 = N_1\,dy_1 \cos\theta_1\,d\theta_1. \tag{5.26}$$

Equations (5.24) through (5.27) can now be combined to give

$$\frac{L_0(\theta_0, \phi)}{N_0^2} = \frac{L_1(\theta_1, \phi)}{N_1^2}, \tag{5.27}$$

which is the general statement of the radiance theorem for an image. This result shows that the image of a plane Lambertian source is also Lambertian and that the basic radiance of the image is equal to that of the source. Thus no optical system can form an image whose basic radiance exceeds that of the object, a result which can alternatively be shown using the laws of thermodynamics.

The discussion in the preceding paragraph showed that the radiance theorem for images could be inferred from the sine condition of Abbe. However, since the radiance theorem can also be deduced from the laws of thermodynamics, it follows that the sine condition must have a thermodynamic interpretation. In fact, R. Clausius [*Pogg. Ann.*, **121**, 1 (1864)] derived the sine condition in 1864, fifteen years before Abbe discovered its importance in the theory of images.

The thermodynamics derivation of the sine condition can be formulated in terms of two small circular blackbodies of equal temperatures T and of areas dA and dA' as shown in Fig. 5.14. Each object is imaged onto the other by a well-corrected optical system. Except for the aperture of the imaging system, each object is surrounded by a perfectly reflecting mirror and thus can change its energy only by interacting with the other. If the radiance of dA measured in its surrounding medium of index N is taken to be $L_0 N^2$, then by the radiance theorem, Eq. (5.22), the radiance of dA' measured in its surrounding medium must be given by $L_0 N'^2$. The flux

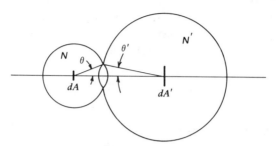

Figure 5.14. Thermodynamic proof of the sine condition.

radiated by dA into an annular element of solid angle of half angle α is then given by

$$d^2\Phi = 2\pi L_0 N^2 \, dA \cos\alpha \sin\alpha \, d\alpha, \qquad (5.28)$$

and thus the total flux transferred from dA to dA' is given by

$$d\Phi = \int_0^\theta d^2\Phi$$

$$= \pi L_0 N^2 \, dA \sin^2\theta, \qquad (5.29)$$

where θ is the half angle subtended by the aperture of the imaging system. Similarly, the flux transferred from dA' to dA is given by

$$d\Phi' = \pi L_0 N'^2 \, dA' \sin^2\theta'. \qquad (5.30)$$

The second law of thermodynamics requires that $d\Phi = d\Phi'$ to prevent a spontaneous heating of one element and a cooling of the other and thus requires that

$$Nh \sin\theta = N'h' \sin\theta', \qquad (5.31)$$

where h and h' are the linear dimensions of the area elements dA and dA', respectively.

5.4 IRRADIANCE

This section considers the calculation of irradiance for a number of circumstances. Two examples were given in Chapter 2. It was shown in Eq.

(2.8) that the irradiance produced by a point source of intensity I on a screen inclined at an angle θ to the line of sight is given by

$$E = \frac{I \cos \theta}{r^2}. \tag{5.32}$$

Thus the irradiance produced by a point source obeys the inverse-square law. Furthermore, the irradiance produced on-axis by a disk Lambertian source of radiance L was shown by Eq. (2.26) to be of the form

$$E = \pi L \sin^2 \theta_{1/2}, \tag{5.33}$$

where $\theta_{1/2}$ is the half angle subtended by the disk at the point of observation.

It is much more difficult to calculate the irradiance produced at an off-axis point by a Lambertian source. A fairly simple solution exists, however, for the special case shown in Fig. 5.15, in which the linear dimensions of the source are much less than the distance z to the plane of observation. Since the distance r between A_s and dA is given by

$$r = \frac{z}{\cos \theta},$$

the solid angle subtended by the area element dA on the observation plane is given by

$$d\Omega = \frac{dA \cos \theta}{r^2}$$

$$= \frac{dA \cos^3 \theta}{z^2},$$

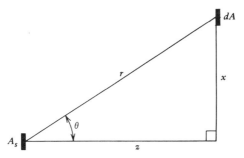

Figure 5.15. Off-axis irradiance produced by a small Lambertian source.

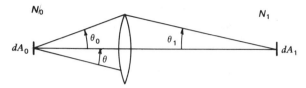

Figure 5.16.   Irradiance of an image.

and the flux radiated from the source of area A_s and radiance L is given by

$$d\Phi = LA_s\cos \theta \, d\Omega$$

$$= \frac{LA_s \, dA \cos^4\theta}{z^2} .$$
(5.34)

Thus the irradiance measured dA is given by

$$E = \frac{d\Phi}{dA}$$

$$= \frac{LA_s}{z^2} \cos^4\theta.$$
(5.35)

This result is sometimes referred to as the cosine-to-the-fourth law. It predicts a rather dramatic decrease in irradiance at off-axis points.

Irradiance of Images

Figure 5.16 shows an area element dA_0 of a Lambertian source of radiance L_0 which is imaged by a well-corrected optical system onto a screen that is placed at normal incidence to the optical axis. The system is assumed to be aplanatic for the object and image points and thus to obey the sine condition of Eq. (5.8). We shall calculate the irradiance of the image by calculating the total flux from dA_0 that is collected by the optical system and then dividing this quantity by the area dA_1 of the image.

The flux radiated by dA_0 into an annular element of solid angle is given by

$$d^2\Phi = L_0 \, dA_0 \cos \theta \, d\Omega$$

$$= 2\pi L_0 \, dA_0 \sin \theta \cos \theta \, d\theta,$$

and thus the flux collected by the optical system becomes

$$d\Phi = \int_0^{\theta_0} 2\pi L_0 \, dA_0 \sin\theta \cos\theta \, d\theta$$

$$= \pi L_0 \, dA_0 \sin^2\theta_0.$$

The image irradiance is then given by

$$E_1 = \frac{d\Phi}{dA} = \pi L_0 \frac{dA_0}{dA_1} \sin^2\theta_0. \tag{5.36}$$

This result can be simplified through use of the sine condition (5.8). If the small quantities h_0 and h, appearing in Eq. (5.8) are taken as the radii of circular area elements dA_0 and dA_1, these area elements must be related by

$$\frac{dA_0}{dA_1} = \frac{N_1^2 \sin\theta_1^2}{N_0^2 \sin\theta_0^2},$$

and Eq. (5.36) for the irradiance can be expressed as

$$E_1 = \pi \frac{L_0}{N_0^2} N_1^2 \sin^2\theta_1. \tag{5.37}$$

The image irradiance at dA_1 can alternatively be calculated by noting that from dA_1 the exit pupil of the optical system appears as a disk of half angle θ_1 and uniform radiance, which is given by the radiance theorem (5.23) as $L_0(N_1^2/N_0^2)$. Equation (5.33) for the irradiance produced by a uniform disk source then gives an expression for the image irradiance that is in agreement with Eq. (5.37).

The quantity $N_1 \sin\theta_1$ appearing in Eq. (5.37) is called the numerical aperture of the imaging system, and the image irradiance is seen to be proportional to the square of this quantity. In geometrical terms, the image irradiance increases with the size of the cone of light converging on dA_1. Another quantity that can be used to describe the speed of an optical system (i.e., the ability of the system to provide a large image irradiance) is the focal ratio, or f-number. The focal ratio of an optical system is usually defined by the relation

$$f^\# = \frac{f}{D}, \tag{5.38}$$

where f is the (second) focal length of the system and D is the diameter of the entrance pupil. The effective focal ratio* $f_{\text{eff}}^{\#}$ is defined by the relation

$$f_{\text{eff}}^{\#} = \frac{1}{2 \tan \theta_1},$$ (5.39)

where θ_1 is the angle defined by Fig. 5.16. For an object at infinity, the effective focal ratio is equal to the focal ratio, although for objects at finite distances the effective focal ratio is larger than the focal ratio. Equation (5.37) for the image irradiance can be expressed in terms of the effective focal ratio in the limit $f_{\text{eff}}^{\#} \gg 1$ as

$$E_1 = \frac{1}{4} \pi L_0 \frac{N_1^2}{N_0^2} \frac{1}{\left(f_{\text{eff}}^{\#} \right)^2}.$$ (5.40)

The Searchlight

Let us consider the irradiance produced by a source placed at a focus of an optical system. The most common example is the searchlight, which is composed of a source, often a carbon arc, placed at the focus of a parabolic reflector.

A schematic searchlight, using lossless, transmitting optics, is shown in Fig. 5.17. A disk Lambertian source of diameter $2r_s$ and radiance L is

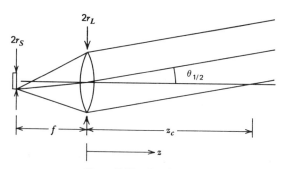

Figure 5.17. Searchlight.

*Some authors define the effective focal ratio as

$$f_{\text{eff}}^{\#} = \frac{1}{2 \sin \theta_1}.$$

Although such a definition is preferred on theoretical grounds, since $\sin \theta_1$ appears in Eq. (5.37), it is the less common definition and thus will not be used here.

placed at the focus of a lens of diameter $2r_L$ and focal length f. Thus light from each point on the surface of the source is collimated by this lens. The beam from the searchlight is seen to diverge at a half angle $\theta_{1/2}$ given by

$$\tan \theta_{1/2} = \frac{r_s}{f}. \tag{5.41}$$

We wish to calculate the irradiance that is produced by the searchlight at any point to the right of the lens. It is useful in this regard to imagine placing one's eye at the point where the irradiance is to be measured and looking back through the lens at the image of the source. This image is then treated as the source of the irradiance, and, as a consequence of the radiance theorem (5.17), it must also have radiance L. Since the image of the source is at infinity, it subtends an angular radius of $\theta_{1/2}$ for all observation points. On the other hand, the angle subtended by the aperture of the lens depends on the observation point, and for points on the optical axis the aperture subtends a half angle given by

$$\tan \phi = \frac{r_L}{z}, \tag{5.42}$$

where z is the distance from the lens to the point of observation. The appearance of the source and lens are shown for several cases in Fig. 5.18. Cases (a)–(c) refer to observation points on the system axis. In (a), the image of the source is smaller than that of the lens, in (b) the image just fills the lens, and in (c) only a portion of the image can be seen through the aperture of the lens. Case (d) shows the appearance from a point off the optical axis. The distance at which the image just fills the lens aperture can be called the critical distance z_c. This distance is calculated by requiring that the angle $\theta_{1/2}$ be equal to the angle ϕ, and thus

$$z_c = \frac{fr_L}{r_s}. \tag{5.43}$$

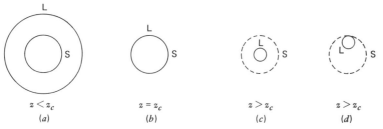

Figure 5.18. Appearance of source (S) and lens (L) from several on-axis observation points (a)–(c) and an off-axis point (d).

This distance can alternatively be inferred from Fig. 5.17 as the distance from the lens to the point where the marginal ray from the edge of the source crosses the optical axis.

The irradiance at normal incidence for axial points at $z < z_c$ can be determined from Eq. (5.34) for the irradiance produced by a disk source of radiance L and half angle $\theta_{1/2}$, the angle subtended by the image of the source:

$$E = \pi L \sin^2\theta_{1/2} = \pi L \frac{r_s^2}{r_s^2 + f^2}$$

or

$$E \simeq \pi L \frac{r_s^2}{f^2} \quad \text{for } z < z_c. \tag{5.44}$$

For axial points at $z > z_c$, the irradiance at normal incidence is that produced by a disk source of radiance L and half angle ϕ, the angle subtended by the lens aperture. By Eq. (5.33), this irradiance is given by

$$E = \pi L \sin^2\phi = \pi L \frac{r_L^2}{r_L^2 + z^2}$$

or

$$E \simeq \pi L \frac{r_L^2}{z^2} \quad \text{for } z > z_c. \tag{5.45}$$

By comparison with Eq. (5.32) for the irradiance produced by a point source, we see that for $z > z_c$ the searchlight can be characterized by an intensity

$$I = L\pi r_L^2. \tag{5.46}$$

The normal irradiance at off-axis points can be determined with the help of Fig. 5.19. In regions I and II, the irradiance does not change with the transverse distance. Thus the irradiance anywhere in region I is given by Eq. (5.44), and the irradiance anywhere in region II is given by Eq. (5.45). In region IV, which lies exterior to the volume bounded by the most transverse rays transmitted by the system, the irradiance is zero. As the point of observation is moved transversely through region III, the irradiance drops monotonically from its on-axis value to zero.

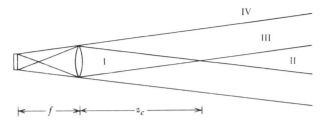

Figure 5.19.   Irradiance produced by a searchlight is given by Eq. (5.44) in region I and by Eq. (5.45) in region II.

5.5   MEASUREMENT OF TOTAL FLUX

Let us consider the problem of determining the total flux transmitted by an optical system. If the system is perfectly transmitting in the sense that its components neither absorb nor reflect and that all rays entering the system exit without encountering any limiting stops or apertures (i.e., there is no vignetting), then it is sufficient to calculate the total flux collected by the system. This flux is given by

$$\Phi = \int \int L(\mathbf{r}, \hat{\mathbf{n}}) \, dA \cos \theta \, d\Omega, \tag{5.47}$$

where $L(\mathbf{r}, \hat{\mathbf{n}})$ is the source radiance at the point r in the direction of the unit vector $\hat{\mathbf{n}}$, where the area integral extends over that portion of the surface of the source which lies within the entrance window and where the angular integral extends over the solid angle subtended by the entrance pupil of the system from the location $\mathbf{r}$.

We wish to characterize the flux-transmitting capabilities of the optical system in a manner that is independent of the properties of the radiation source. If we assume first that the source is uniform and Lambertian with radiance L_0, Eq. (5.47) becomes

$$\Phi = L_0 \int \int dA \cos \theta \, d\Omega, \tag{5.48}$$

which can be expressed as

$$\Phi = \frac{L_0}{N_0^2} \mathcal{E} \tag{5.49}$$

where the *étendue* of the system is defined as

$$\mathcal{E} = N_0^2 \int \int dA \cos \theta \, d\Omega. \tag{5.50}$$

Thus the étendue is a purely geometrical quantity and is a measure of the flux-gathering abilities of the system in that the collected flux is given by the product of this quantity with the basic radiance of the source.

Some intuition regarding the significance of the étendue can be obtained by considering the case of a source whose area A_0 is sufficiently small that the solid angle subtended by the entrance pupil does not change appreciably over the dimensions of the source. Figure 5.16 describes this situation if dA_0 and dA_1 are replaced by finite sources of size A_0 and A_1, respectively. Then expression (5.50) for the étendue becomes

$$\mathcal{E} = N_0^2 A_0 \int \cos \theta \, d\Omega \tag{5.51}$$

or

$$\mathcal{E} = N_0^2 A_0 \Omega_{\text{proj},0}, \tag{5.52}$$

where the projected solid angle of the source is defined by

$$\Omega_{\text{proj},0} = \int \cos \theta \, d\Omega, \tag{5.53}$$

with the integration extending over the entrance pupil of the source. Since the entrance pupil subtends a half angle θ_0, the projected solid angle is given by

$$\Omega_{\text{proj},0} = \int_0^{\theta_0} 2\pi \sin \theta \cos \theta \, d\theta$$

$$= \pi \sin^2\theta_0, \tag{5.54}$$

and by Eq. (5.52) the étendue becomes

$$\mathcal{E} = \pi N_0^2 A_0 \sin^2\theta_0. \tag{5.55}$$

This expression is equal (within factors of π) to the square of the quantity $Nh \sin \theta$, which by Abbe's sine condition (5.8) is invariant between image and object points for a well-corrected imaging system. Thus the étendue is seen to be invariant between the object and image planes.

This conservation property of the étendue can be formulated in a much more general way. By differentiating Eq. (5.50), an expression for an element of étendue can be obtained:

$$d^2\mathcal{E} = N^2 \, dA \cos \theta \, d\Omega. \tag{5.56}$$

The quantities appearing in this equation are illustrated in Fig. (5.20), where θ is the angle between the normal to dA and the element $d\Omega$ of solid angle. This element $d^2\mathcal{E}$ of étendue can be associated with the ray that passes through dA in the direction of $d\Omega$. As this ray is followed through the optical system, an element of étendue can be determined using a similar procedure for each point along the ray. We shall now show that this element $d^2\mathcal{E}$ of étendue is invariant along this ray. The flux leaving dA in the direction of $d\Omega$ is given by

$$d^2\Phi = L \, dA \cos \theta \, d\Omega, \tag{5.57}$$

which can be expressed by Eq. (5.56) as

$$d^2\Phi = \frac{L}{N^2} d^2\mathcal{E}. \tag{5.58}$$

If the optical system is lossless, $d^2\Phi$ must be conserved according to the law of conservation of energy. In addition, the basic radiance L/N^2 was previously shown to be an invariant. Therefore, $d^2\mathcal{E}$ also must be invariant along the ray associated with $d^2\mathcal{E}$. Since the total étendue of the system is given by

$$\mathcal{E} = \int \int d^2\mathcal{E}, \tag{5.59}$$

it is also a conserved quantity. This implies that the same value of the étendue will be obtained if Eq. (5.59) is evaluated with the area integral performed over any reference surface that intersects all the rays passing through the system.

Integrating Sphere

It is often necessary to measure the flux in a beam of nonuniform irradiance. For sufficiently small beams, a detector of large active area can

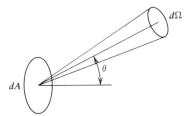

Figure 5.20. Geometry of an element of étendue.

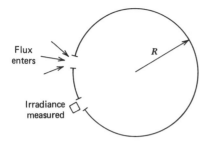

Figure 5.21.   Integrating sphere.

be used to do this. For larger beams, or when the response of the detector across its active area is not sufficiently uniform, an integrating sphere can be used to produce a uniform irradiance that is proportional to the total incident flux. This irradiance is then measured by a detector, as shown in Fig. 5.21.

The interior of the integrating sphere is coated with a material of highly diffuse reflectivity. Thus incident radiation will be scattered many times before being absorbed, and the radiation will become very uniformly distributed over the interior surface of the sphere. An integrating sphere thus produces a uniform irradiance that is proportional to the total flux entering the sphere.

The theoretical basis for understanding the operation of an integrating sphere is shown in Fig. 5.22. We wish to calculate the contribution to the irradiance at dA' due to the source element dA, which is assumed to be a Lambertian source of radiance L. Due to the spherical geometry of the problem, the angles θ and θ' defined in the figure are equal. The flux transmitted from dA to dA' is given according to the definition of radiance as $d^2\Phi = L\, dA_{\text{proj}}\, d\Omega$, where the projected source area is given by $dA \cos\theta$ and the solid angle subtended by dA' is given by $dA' \cos\theta/d^2$. Since the

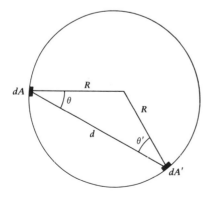

Figure 5.22.   Calculation of the irradiance on the inner surface of an integrating sphere.

distance d is equal to $2R\cos\theta$, the flux can be expressed as $d^2\Phi = L\,dA\,dA'/4R^2$, and the contribution to the irradiance at dA' is given by

$$dE = \frac{d^2\Phi}{dA'} = \frac{L\,dA}{4R^2}.$$ (5.60)

We note that this result is independent of the angle θ. Thus the flux leaving each part of the spherical surface is distributed uniformly over the remainder of the surface. If an amount of flux Φ enters the sphere, the irradiance at any point on the sphere due to a single reflection will be given by $\Phi r/4\pi R^2$, where r is the diffuse reflectivity of the surface. A fraction r of this flux will be reflected a second time, leading again to a uniform distribution over the surface of the sphere. The total surface irradiance is thus given by

$$E = \frac{\Phi}{4\pi R^2}\left(r + r^2 + r^3 + \cdots\right)$$

$$= \frac{\Phi}{4\pi R^2}\frac{r}{1 - r}.$$ (5.61)

BIBLIOGRAPHY

M. Born and E. Wolf, *Principles of Optics*, Pergamon, New York, 1975, Chapters III–V.

P. Drude, *The Theory of Optics*, Dover, New York, 1959, Part III.

F. A. Jenkins and H. E. White, *Fundamentals of Optics*, McGraw-Hill, New York, 1957, Part I.

R. Kingslake, "Basic Geometrical Optics," in *Applied Optics and Optical Engineering*, R. Kingslake, ed., Academic New York, 1965.

M. V. Klein, *Optics*, Wiley, New York, 1970, Chapters 3 and 4.

See also the works listed in the bibliography for Chapter 2.

PROBLEMS

(Many interesting radiometric calculations are better stated in photometric terms, and some examples of these are presented at the conclusion of Chapter 6.)

1 Prove the results given by the numbered equations of Section 5.1, all of which were stated without proof.

2 A large Lambertian source is immersed in a medium of refractive index $n = 1.414$. The source radiance measure in this medium is 1.0 W/m^2 sr.

What is the irradiance measured at point a located 4 m from a 6-m-diameter opening in screen S?

3   A photographer photographs a landscape 5 km away on a sunny day at an exposure of $f/10$ at $\frac{1}{100}$ sec. Using the same film and the same f-number, what exposure time should be used to photograph the moon? You may assume that the diffuse reflectivity of the moon and the landscape are similar. (This last assumption is not entirely realistic, but it can be used to determine an exposure time that will provide a usable photograph).

6

Photometry and Vision

In this chapter we introduce the study of photometry, that is, the study of the transfer of visible light through optical systems. Photometry differs from radiometry, which we have discussed in Chapters 2 through 5, in that photometry is concerned only with that part of the radiation field which can induce a visual response. In fact, photometry is concerned with quantifying the visual response that can be produced by the radiation. The most fundamental quantity of photometry is the luminous flux Φ_v; it is the analog of the radiant flux Φ of radiometry and may be considered to be the integral of the spectral radiant flux Φ_λ over the visible spectrum with each wavelength interval weighted in proportion to its visibility. We shall consistently follow the convention of designating the photometric quantities by the same symbol used for the analogous radiometric quantity, but followed by a subscript v. This notation will serve as a reminder that the theory of photometry is formally identical to that of radiometry and thus that the formal properties of radiometry derived earlier apply equally well to photometry. It is important to realize, however, that rather different properties of the radiation field are being measured in these two theories. In radiometry, the property being measured is the energy of the field; in photometry, it is the visual response produced by the field.

6.1 PHOTOMETRIC QUANTITIES AND UNITS

As discussed earlier, photometric quantities can be defined by analogy to their radiometric counterparts, and the quantities so defined are listed in Table 6.1. The definitions given in the table are meant to be heuristic; the precise definitions of the radiometric equivalents of these quantities have been given in Chapter 2. The units of the photometric quantities are also given in Table 6.1 in the system based on the lumen (or the candela, which is 1 lumen per steradian) and the meter, second, and steradian. Several additional units sometimes employed in photometry are defined for purpo-

Table 6.1. Photometric Quantities and Units

Quantity	Symbol	Definition	Units[a]
Luminous energy	Q_v	$\int \Phi_v \, dt$	lm s = talbot
Luminous density	U_v	dQ_v/dV	lm s/m^3
Luminous flux	Φ_v	dQ_v/dt	lm
Illuminance	E_v	$d\Phi_v/dA$	lm/m^2 = lx
Luminous exitance	M_v	$d\Phi_v/dA$	lm/m^2 = lx
Luminance	L_v	$d^2\Phi_v/dA_{proj}\, d\Omega$	lm/m^2 sr = nt
Luminous intensity	I_v	$d\Phi_v/d\Omega$	lm/sr = cd

[a] The candela is abbreviated cd; the lumen is abbreviated lm; the lux is abbreviated lx; the nit is abbreviated nt; there is no standard abbreviation for the talbot.

ses of reference in Table 6.2. The units of lamberts, foot-lamberts, and apostilbs are conventionally applied only to Lambertian sources.

Since all of photometric quantities of Table 6.1 can be measured in units based on the lumen, meter, second, and steradian, an absolute definition of any one of the photometric units serves to fix all the units. Historically, a unit of intensity has provided the absolute standard for the photometric units. Initially, the luminous intensity of a candle of particular construction was taken as the definition of the standard candle. An improvement was made when the definition of the *international candle* was adopted as a fixed fraction of the luminous intensity of a standardized electric lamp operated under prescribed conditions. Independently, a flame standard known as the

Table 6.2. Photometric Units

Units of Illuminance

foot-candle (fc)	$\equiv$ lm/ft^2
lux (lx)	$\equiv$ lm/m^2
phot (ph)	$\equiv$ lm/cm^2

Units of Luminance

apostilb (asb)	$\equiv \dfrac{1}{\pi}$ cd/m^2
foot-lambert (fL)	$\equiv \dfrac{1}{\pi}$ cd/ft^2
lambert (L)	$\equiv \dfrac{1}{\pi}$ cd/cm^2
nit (nt)	$\equiv$ cd/m^2
stilb (sb)	$\equiv$ cd/cm^2

Table 6.3. Illuminance Levels Produced by Several Sources

Source	Illuminance
Sun at zenith	1.2×10^5 lux (lm/m^2)
Clear sky	1×10^4
Overcast sky	1×10^3
60-watt incandescent lamp at 1 m	1×10^2
Candela at 1 m	1
Full moon at zenith	0.27
Moonless overcast sky	1×10^{-4}

Hefner candle was introduced, which has an intensity of 0.9 international candle. These standards suffered from the problem that their intensities depended on the details of their physical construction. A more absolute definition was thus adopted in 1939, when the *new candle* was defined as 1/60 the luminous intensity of a 1-cm^2 blackbody radiator at the temperature of the freezing point of platinum (roughly 2045 K). The new candle as so defined was determined to be about 2% smaller than the international candle. Since 1948, the term *candela* (abbreviated cd) has replaced the term new candle in order to avoid any possible confusion regarding this 2% discrepancy.

In 1977, a decision was made by the International Committee for Weights and Measures to choose a unit of luminous flux (rather than a unit of luminous intensity) as the primary photometric standard, and to define the lumen as the luminous flux of monochromatic radiation at 555 nm whose radiant flux is equal to 1/683 W. This new definition is consistent with the former definition and lies within the error limits of the former definition imposed by the uncertainty in the freezing point of platinum. A

Table 6.4. The Luminance of Several Sources

Source	Luminance
Surface of sun	2×10^5 cd/cm^2
Carbon arc lamp	1×10^4
750-watt projection lamp filament	2×10^3
60-watt frosted incandescent lamp	9
CRT	5
40-watt fluorescent lamp	0.5
Overcast sky	0.2
Self-luminous paint	3×10^{-6}

wavelength of 555 nm was chosen because the light-adapted human eye is most sensitive at this wavelength. The definition of the lumen applies, however, to vision with the eye in any state of light or dark adaptation.

In order to provide some feeling for the size of the lumen and the candela, Table 6.3 lists the illuminance levels encountered in a number of common circumstances, and Table 6.4 lists the luminances of a number of common light sources. Examination of these tables can provide a feeling for the expected magnitudes of these photometric quantities.

6.2 HUMAN VISION

A cross-sectional view of a human right eye is shown in Fig. 6.1. Most of the optical power of the eye is due to refraction at the interface between the cornea and air. The optical power of the eye lens is controlled by muscles that can control its shape, allowing the eye to accommodate various object distances. Within the limits of accommodation, an image is formed on the retina, which is the light-sensitive part of the eye. The region of the retina that produces the greatest visual accuity is called the fovea; this region corresponds to a visual field whose diameter is approximately 1°. Conversely, the blind spot corresponds to the region on the retina where no light sensors are located because this is where the optic nerve is attached. The vitreous humor is the material contained between the retina and the lens,

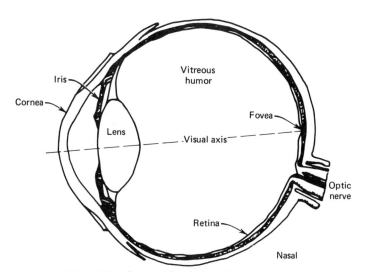

Figure 6.1. Cross-sectional view of a human right eye.

and has a refractive index of $N = 1.336$. As such, the typical 17-mm focal length of the relaxed eye in the object space differs from the typical 22-mm focal length of the eye in the image space. These focal lengths imply that an object subtending an angle α produces an image of linear size 17α mm on the retina. The amount of light entering the eye is controlled by the iris, which can vary in diameter from 2 to 8 mm.

The apparent brightness of an object depends most directly on the illuminance of the retina caused by the object. The retinal illuminance E_v is related by Eq. (5.37) to the luminance L of the object being viewed by

$$E_v = \pi L_v N^2 \sin^2\theta_{1/2}, \tag{6.1}$$

where $N = 1.336$ is the refractive index of the vitreous humor, and where $\theta_{1/2}$ is the half angle subtended by the pupil at the retina. Thus the retinal illuminance is proportional to the source luminance (explaining why this quantity is sometimes called the brightness) and to the square of the numerical aperture of the eye.

The retina of the eye contains two types of visual receptors called rods and cones. Cones are primarily responsible for vision at high luminance levels, whereas rods are primarily responsible for vision at low luminance levels. Since only the cones contribute to color vision, a complete loss of color sensation occurs at low light levels.

The visual system can adapt to a wide range of luminance levels. Adaptation to luminance levels of the visual field of $\geqslant 3$ cd/m^2 leads to light-adapted or photopic vision, whereas adaptation to field luminances of $\leqslant 3 \times 10^{-5}$ cd/m^2 leads to dark-adapted or scotopic vision. Mesopic vision occurs if the visual system is adapted to a field luminance between these two levels. An initially dark-adapted eye becomes light-adapted in 2 to 3 minutes of exposure to high luminance levels, whereas approximately 45 minutes are required for an initially light-adapted eye to become dark-adapted.

Since only cones are located in the fovea of the retina, vision at low luminance levels is enhanced by fixating slightly to the side of an object to be viewed. For radiation in the range 500 to 550 nm, the visual response near the periphery of a dark-adapted eye is 300 times greater than that at the fovea.

Spectral Response

It is possible to measure the flux of a beam of light in either photometric or radiometric units. In order to establish the relationship between these two quantities, it is necessary to determine the relative efficiency of radiation of different wavelengths in producing a visual response. The accepted results of such a determination are shown in Fig. 6.2, where the relative sensitivity of the eye at various wavelengths is denoted as the *spectral luminous efficiency* functions V_λ and V'_λ. The curve V_λ applies to light-adapted (photopic) vision, whereas the curve V'_λ applies to dark-adapted (scotopic) vision. The shift of the maximum of the spectral luminous efficiency curve toward blue wavelengths as the eye dark adapts is known as the Purkinje effect.

One method of determining the spectral luminous efficiency function V is through what is known as the direct method. Here an observer is shown two adjacent areas illuminated by monochromatic radiation of wavelengths λ and λ_0 and is allowed to vary the radiance of the region of wavelength λ until its apparent brightness equals that of the region of wavelength λ_0. This process is repeated for different values of λ, and the relative luminous efficiency is given by the inverse of the radiance required to produce a constant apparent brightness. This method is conceptually straightforward, but has the disadvantage that many observers find it difficult to compare the apparent brightnesses of regions of different color.

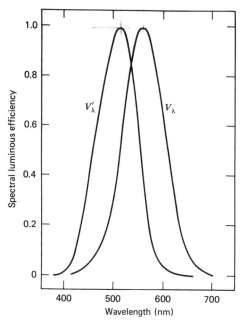

Figure 6.2. The relative sensitivity of human vision to monochromatic radiation for light-adapted (V_λ) and dark-adapted (V'_λ) vision.

This difficulty is overcome by the step-by-step method. In this method, an observer is shown two adjacent areas illuminated by monochromatic radiation of wavelengths λ and $\lambda + \Delta\lambda$ and is asked to vary the radiance of one of them until their apparent brightnesses are equal. The relative radiances of the two regions are then recorded, and the observer is asked to repeat the procedure with the regions illuminated at wavelengths $\lambda + \Delta\lambda$ and $\lambda + 2\Delta\lambda$. The relative luminous efficiency is then determined as the inverse of the spectral radiance required to produce a constant apparent brightness. This method can be used to trace out the spectral luminous efficiency curve over the entire visible region. Since the observer compares the apparent brightness of areas of nearly the same color, good accuracy can be obtained.

The method most generally used is the flicker method. In this method, an observer is presented with a sequence of flashes consisting of white light alternating with monochromatic light of variable radiance. A flicker rate of from 10 to 20 times per second is chosen, and the subject is asked to vary the radiance of the monochromatic light so that the flicker of the combined signal is minimized. The spectral luminous efficiency is then taken to be proportional to the inverse of this radiance. Although this method seems artificial, it gives good agreement with the results obtained using other methods and is often easier to implement.

No matter which method is used, the results for a large number of observers are averaged to obtain the luminous efficiency curve for a hypothetical standard observer.

The relation between the photometric and radiometric quantities can be obtained as follows: The *spectral luminous efficacy* K_λ for photopic vision is defined as

$$K_\lambda = \frac{\Phi_{v,\lambda}\, d\lambda}{\Phi_\lambda\, d\lambda} \tag{6.2}$$

(in lumens per watt), where $\Phi_{v,\lambda}\, d\lambda$ is the element of luminous flux corresponding to the element $\Phi_\lambda\, d\lambda$ of radiant flux. According to the definition of the lumen given earlier, K_λ has the value

$$K_m = 683 \text{ lm/W} \tag{6.3}$$

at a wavelength of 555 nm, which is the wavelength of maximum sensitivity for photopic vision. At other wavelengths, K_λ is given by

$$K_\lambda = K_m V_\lambda, \tag{6.4}$$

since the spectral luminous efficiency function V_λ shown in Fig. 6.2 gives the relative value of the visual response for light of wavelength λ. Abney's law states that visual response is additive in the sense that the luminous flux Φ_v produced by a continuous spectrum is given by the relation

$$\Phi_v = K_m \int V_\lambda \Phi_\lambda \, d\lambda, \qquad (6.5)$$

where Φ_λ is the spectral radiant flux of the signal. Abney's law is in fact an approximation in that deviations from the predictions of Eq. (6.5) of as large as 40% have been reported if the visual response to a continuous spectrum is compared directly to the response to monochromatic light. Interestingly, no deviations from Abney's law are found if the comparison is made by the flicker method, and thus determinations of the spectral luminous efficiency function V_λ using the flicker method are the more suitable for use in Eq. (6.5).

Photopic vision is often assumed in photometric calculations unless the converse is explicitly stated.

For the case of scotopic or dark-adapted vision, analogous relations can be obtained. The spectral luminous efficacy is now denoted by K'_λ and is related to the spectral luminous efficiency V'_λ by

$$K'_\lambda = K'_m V'_\lambda. \qquad (6.6)$$

Since the definition of the lumen states that K'_λ is equal to 683 lm/W at 555 nm, where by Fig. 6.2 V'_λ has the value 0.40, the constant $K_{m'}$ must have the value 1700 lm/W. The spectral luminous efficiency function V'_λ reaches this maximum value at a wavelength of 510 nm.

In order to compare the eye to the electronic radiation detectors to be discussed in the following chapters, we shall quote here some of the characteristics of the eye as a radiation detector. The angular resolution of the human eye is approximately 1 minute of arc, and the temporal resolution of the eye is approximately 0.02 second. Another important attribute of the eye is its ability to detect weak optical signals. It has been found that for sufficiently large objects, the threshold luminance is approximately 10^{-3} cd/m^2 for photopic vision and 10^{-6} cd/m^2 for scotopic vision. A precise determination of the minimum radiant energy required to produce a visual response has been performed by S. Hecht, S. Shaer, and M. H. Pirenne [*J. Gen. Physiol.*, **25**, 819 (1942)]. They report a minimum threshold energy (i.e., the energy incident on the eye that produces a visual response with 60% probability) for dark-adapted vision of $(3.9 \pm 1.8) \times 10^{-17}$ J for excitation at 510 nm, using a flash of 1-ms duration and a 10 arc minute field of view. This energy corresponds to between 50 and 150 photons incident on the cornea. There is some evidence that at most 10% of this light is absorbed by the visual receptors. Thus only a very small number of photons need be involved in the actual detection process.

BIBLIOGRAPHY

T. N. Cornsweet, *Visual Perception*, Academic, New York, 1970.

Glen A. Fry, "The Eye and Vision," in *Applied Optics and Optical Engineering*, Vol. 2, R. Kingslake, ed., Academic, New York, 1965.

C. H. Graham, ed., *Vision and Visual Perception*, Wiley, New York, 1965.

Illumination Engineering Society, *I. E. S. Lighting Handbook*, 4th ed., New York, 1966.

R. Kingslake, "Illumination of Optical Images," in *Applied Optics and Optical Engineering*, Vol. 2, R. Kingslake, ed., Academic, New York, 1965.

Y. LeGrand, *Light, Color and Vision*, Chapman and Hall, London, 1968.

RCA, *Electro-Optics Handbook*, Harrison, New Jersey, 1974.

R. P. Teele, "Photometry" in *Applied Optics and Optical Engineering*, Vol. 1, R. Kingslake, ed., Academic, New York, 1965.

J. W. T. Walsh, *Photometry*, Constable and Company, London, 1953.

PROBLEMS

1 Derive an expression for the luminance of a blackbody of temperature T. Sketch the form of this function. Evaluate this function to an accuracy of 10% for temperatures of 2000, 6000, and 20,000 K. (Assume photopic vision).

2 Using the results of Problem 1, determine the luminous efficacy of a blackbody of temperatures 2000, 6000, and 20,000 K.

3 Estimate the minimum temperature at which a blackbody radiator becomes visible for dark-adapted vision.

4 (a) Calculate the illuminance at point a located 100 m from a 100-million-candlepower searchlight utilizing a 1-m diameter $f/4$ reflector and a disk source of radius 1 cm.

   (b) What is the distance between points a and b where point b is the most distant transverse point where the illuminance equals that at point a?

   (c) What is the distance between points a and c, where point c is the nearest transverse point where the illuminance is zero?

(d)   Calculate the illuminance at point d located 1.0 km from the searchlight.

(e)   What is the distance between points d and e where point e is the nearest transverse point where the illuminance is zero?

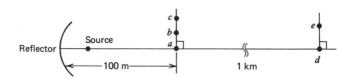

5   The Köhler method of projecting an image from photographic film onto a screen is illustrated in the following figure. The illumination is provided by a lamp of luminance L_v and area A, which is imaged by the condenser lens onto the projection lens. The transparency, whose transmittance is T, is imaged onto the screen by the projection lens. These lenses have focal lengths f_c and f_p and focal ratios $f_c^{\#}$ and $f_p^{\#}$, respectively.

(a)   Derive an expression for the screen illuminance.

(b)   Evaluate this expression numerically for a 1-cm^2 tungsten source that you may assume emits like a blackbody of temperature 2800 K. Assume that $s_1 = 25$ mm, $s_2 = 100$ mm, and $s_3 = 5$ m.

(c)   What role do the focal ratios of the lenses play in determining the photometric properties of the projector?

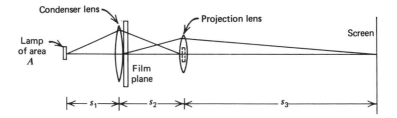

6   Using the concept of retinal illuminance, explain why stars can be seen during daylight hours through a large telescope, but not with the unaided eye. Please include explicit formulas for the retinal illuminance produced by the star and by the sky background.

7   What is the apparent diameter of the sun as measured at the bottom of a
    swimming pool? What is its luminance? What is the illuminance pro-
    duced on a surface that is beneath the surface of the water and
    perpendicular to the line of sight to the sun?

    Assume instead that the sky is overcast and has a uniform luminance
    of L. What is the illuminance on the bottom of the pool, and how does it
    compare to that on the ground?

7

Radiation Detectors

This chapter begins a discussion of the detection of radiation. The main purpose of this chapter is to introduce some of the terminology used to describe detectors and to introduce some concepts that will be described in more detail in the following chapters.

We shall consider a radiation detector to be a device that produces an output signal which depends on the amount of radiation hitting the active region of the detector. This output is usually an electrical signal, but it can be a mechanical deflection of some element or, in the case of detection by vision, a physiological response.

7.1 CLASSES OF DETECTORS

It is often convenient to classify detectors as belonging to one of three major types.

Thermal Detectors

A thermal detector converts incident radiation into heat, thereby raising the temperature of some element of the detector. This change in temperature is then converted to an electrical (or mechanical) signal that can be amplified and displayed.

Thermal detectors have the desirable property that they are often capable of responding to a wide range of wavelengths without appreciable variations in sensitivity. They are used when a broad spectral range is necessary, and sometimes they are used to calibrate other detectors.

Several examples of thermal detectors are shown in Fig. 7.1. In the radiation thermocouple shown in Fig. 7.1a, radiation is allowed to fall onto one of two identical targets, thereby causing the targets to come to different temperatures. Thermocouples attached to each target and wired in series thus produce by the Seabeck effect an output voltage that is proportional to the radiant power hitting the active target.

The bolometer, shown in Fig. 7.1b, is a thermal detector that produces an electrical signal by the temperature-dependent change in electrical resistance of its active element. The resistance of many metals changes with temperature according to

$$\frac{\Delta R}{R} = \alpha \, \Delta T$$

where α is typically of the order $0.003/°C$.

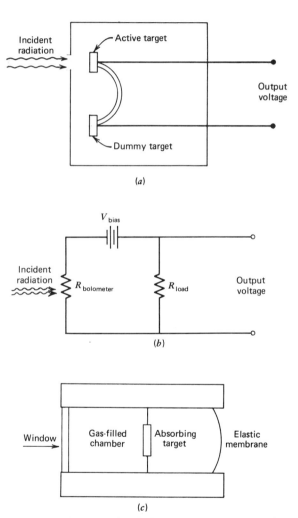

(a)

(b)

(c)

Figure 7.1. Representative detectors of optical radiation. (a) radiation thermocouple; (b) bolometer; (c) Golay cell.

The Golay cell [M. J. E. Golay, *Rev. Sci. Instrum.*, **18**, 346 (1947); **18**, 357 (1947); **20**, 816 (1949)], shown in Fig. 7.1c, contains a target that absorbs incident radiation, thereby raising the temperature of the enclosed gas. A membrane that acts as a wall of the enclosure is thus distorted. The reflection of a beam of visible light from the membrane is used to monitor the motion of the membrane.

Photon Detectors

Photon (or quantum) detectors respond to incident radiation without first thermalizing its energy. The detection process involves a change in the characteristics of the detector caused by the absorption of individual photons. Examples include the emission of an electron from the surface of a metal (the external photoelectric effect) and the production of an electron-hole pair in the interior of a semiconductor (the internal photoelectric effect). Examples of photon detectors include the vacuum photodiode, the photomultiplier, and photoconductive and photovoltaic solid-state detectors.

The vacuum photodiode is shown schematically in Fig. 7.2. Incident radiation causes photoelectrons to be ejected from the photocathode. An applied bias voltage causes this charge to be collected at the anode, giving rise to a photocurrent that is measured by the external circuit.

Coherent Detectors

Coherent detectors produce an output signal that is a measure of the electric field strength of the optical signal, thus preserving the phase information present in the optical field. An example is the heterodyne

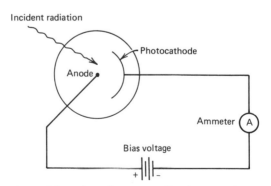

Figure 7.2. Schematic vacuum photodiode circuit.

receiver, in which the optical signal is mixed with a monochromatic "local-oscillator" laser field to produce a beat note at radio frequencies. The optical signal is thereby shifted to radio frequencies, where a variety of signal-processing operations can be performed.

The use of coherent detection systems at optical frequencies is somewhat uncommon, but the use is warranted when good frequency resolution or the retention of phase information is required.

7.2 CHARACTERIZATION OF DETECTORS

This section defines some of the commonly used descriptions of detectors.

The responsivity of a detector is defined most generally as

$$\mathcal{R} = \frac{\text{output signal}}{\text{input power}}. \tag{7.1}$$

The units of the responsivity depend on the form of the electrical response exhibited by a given detector; thus the responsivity of a detector that produces an output voltage proportional to the incident power can be measured in volts per watt.

In general, the responsivity will depend on the wavelength of the incident radiation, and thus the spectral response of a detector can be specified in terms of its responsivity as a function of wavelength.

A common technique in detection is to modulate the radiation beam to be detected and to measure only the modulated component of the electrical output of the detector, as shown in Fig. 7.3. This technique avoids the problem of baseline drifts that often affects electronic amplifiers, since only the ac component of the electrical signal need be measured. In addition, this technique can provide some discrimination against electrical noise, since the signal is contained only in the Fourier component of the electrical signal at the modulation frequency, whereas the electrical noise is often broadband.

An important characteristic of a detector is therefore its frequency response, that is, its ability to respond to rapidly modulated radiation. The response to radiation modulated at frequency f is defined by

$$\mathcal{R}(f) = \frac{v_S(f)}{P_S(f)}, \tag{7.2}$$

where $P_S(f)$ is the rms value of the signal power contained within the harmonic component at frequency f and $v_S(f)$ is the rms output voltage

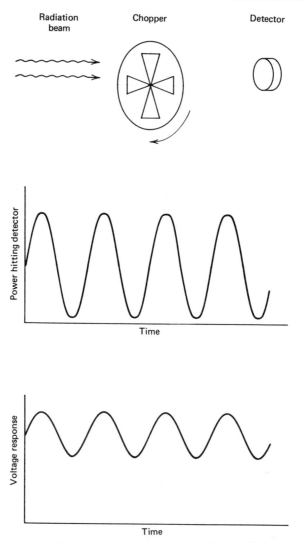

Figure 7.3. Detection of temporally modulated radiation.

within this same harmonic component. In general, the responsivity $\mathcal{R}(f)$ of a detector to radiation modulated at frequency f decreases for large modulation frequencies. In many practical situations, the frequency response of a detector can be characterized by a single time constant τ. (Even if this is not the case, one can imagine filtering the detector's output through a low-pass filter of time constant τ, in which case the ensuing analysis holds). This implies that if a delta-function pulse of radiation of the form

$$P(t) = Q_0 \delta(t)$$ (7.3)

is incident on the detector, an output voltage signal of the form

$$v(t) = \begin{cases} 0 & t < 0 \\ v_0 e^{-t/\tau} & t \geq 0 \end{cases}$$ (7.4)

is produced. Since a delta-function input pulse contains all frequency components in equal amounts, the frequency response $\mathcal{R}(f)$ must be proportional to the Fourier transform of the impulse response function $v(t)$, which is given by

$$V(f) = \int_{-\infty}^{\infty} v(t) e^{-i2\pi ft} dt$$

$$= \frac{v_0 \tau}{1 + i2\pi f\tau}.$$ (7.5)

The frequency response is thus of the form

$$\mathcal{R}(f) = \frac{\mathcal{R}_0}{1 + i2\pi f\tau},$$ (7.6)

where the dc responsivity has been denoted by $\mathcal{R}_0 = v_0\tau/Q_0$. The frequency response is seen to be complex, implying the existence of a phase shift between the input and output waveforms. In many circumstances, only the magnitude of the voltage response is of concern, and this is given by the modulus of $\mathcal{R}(f)$ as

$$|\mathcal{R}(f)| = \sqrt{\mathcal{R}(f)\mathcal{R}(f)^*} = \frac{\mathcal{R}_0}{\left[1 + (2\pi f\tau)^2\right]^{1/2}}.$$ (7.7)

This function is plotted in Fig. 7.4.

It is useful to characterize the frequency response shown in Fig. 7.4 by its cutoff frequency, defined as the modulation frequency f_c at which $|\mathcal{R}(f_c)|^2$ falls to one-half its maximum value. The cutoff frequency is then related to the response time by

$$f_c = \frac{1}{2\pi\tau}.$$ (7.8)

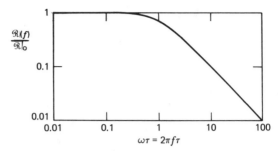

Figure 7.4. Frequency response of a detector characterized by an exponential decay time τ.

Another common descriptor of detection systems is the rise (or fall) time of the electrical response. If the power incident on the detector is in the form of the step function

$$P(t) = \begin{cases} 0 & t < 0 \\ P_0 & t \geqslant 0 \end{cases} \tag{7.9}$$

and if the impulse response function of the system is given by Eq. (7.4), the electrical response of the detection system will be of the form

$$v(t) = \begin{cases} 0 & t < 0 \\ \mathcal{R}_0 P_0 (1 - e^{-t/\tau}) & t \geqslant 0. \end{cases} \tag{7.10}$$

One common definition of the rise time of the output pulse is the time interval required for the output signal to rise from 10 to 90% of its final value. Evaluation of Eq. (7.10) using this criterion shows that the rise time T is related to the response time τ by

$$T = 2.20\tau.$$

The rise time of such a system is thus related to its cutoff frequency by

$$T = \frac{0.35}{f_c}.$$

7.3 NOISE IN THE DETECTION PROCESS

Our ability to detect small amounts of radiant energy is inhibited by the presence of noise in the detection process. By noise, we mean random fluctuations in the electrical signal from the detector; thus we explicitly exclude spurious signals from a known source, such as a 60-Hz pickup. These spurious signals, being nonrandom, can in principle be eliminated.

One source of noise can be found in the intrinsic fluctuations of the radiation field itself. For the case of thermal radiation, these fluctuations have already been discussed in connection with Eq. (3.62). The name photon noise is often given to the noise resulting from these fluctuations. A detection system in which photon noise is much larger than other sources of noise can be considered to be ideal in the sense that it is impossible to improve the performance of the system. Such a detection system is often said to be photon-noise limited. Photon noise can be considered to be the origin of shot noise in photon detectors and of temperature noise (fluctuations in the temperature of the active element) in thermal detectors. These noise sources and their interrelations will be discussed in detail in the following chapters.

In determining the photon-noise-limited performance of detectors, it is important to distinguish between noise in the signal to be detected and noise in any radiation background that is present even in the absence of the signal. This distinction is particularly important at infrared wavelengths, where thermal emission from the atmosphere as well as from optical components is likely to create a large radiation background.

Other sources of noise are less fundamental in that they result from noise generated within the detector or in its associated electronics. Examples include the fluctuations in the dark current of a photomultiplier and Johnson noise (i.e., thermally induced voltage fluctuation in resistors).

Since noise produces a random fluctuation in the output of a radiation detector, it can mask the output that is produced by a weak optical signal. Noise thus sets limits on the minimum input power that can be detected under given conditions. One convenient description of this minimum detectable signal power is the noise equivalent power (NEP), which is defined to be the input power that produces an output signal-to-noise ratio of unity in a specified electrical bandwidth Δf. The electrical bandwidth, which is defined more precisely later, is approximately given by the inverse of the response time of the detection system. It is often implicitly assumed that the NEP is referred to a 1-Hz electrical bandwidth, and when this assumption is true the NEP may be interpreted as approximately the minimum signal power that can be detected in a measurement of 1-second duration.

Let v_N denote the rms noise voltage produced by a radiation detection system whose modulation frequency is f and whose electrical bandwidth is Δf. The NEP of this system is then given by the expression

$$P_N(f, \Delta f) = \frac{v_N}{|\mathcal{R}(f)|}, \tag{7.11}$$

where $\mathcal{R}(f)$ is the responsivity defined by Eq. (7.2). The NEP is usually quoted for a fixed reference bandwidth $(\Delta f)_{\text{ref}}$, which is often assumed to be 1 Hz. Since for most noise sources v_N is proportional to $\sqrt{\Delta f}$, the value of the NEP for other bandwidths can then be calculated from the relation

$$P_N(f, \Delta f) = P_N[f, (\Delta f)_{\text{ref}}] \sqrt{\frac{\Delta f}{(\Delta f)_{\text{ref}}}}. \tag{7.12}$$

Thus an optical signal modulated harmonically at frequency f and containing rms power $P_S(f)$ in this harmonic component will produce an rms signal voltage given by $v_S = |\mathcal{R}(f)|P_S(f)$. The signal-to-noise ratio of the output signal is thus given by

$$\frac{S}{N} \equiv \frac{v_S}{v_N} = \frac{P_S(f)}{P_N(f, \Delta f)} = \frac{P_S(f)}{P_N(f, (\Delta f)_{\text{ref}})} \sqrt{\frac{(\Delta f)_{\text{ref}}}{\Delta f}}. \tag{7.13}$$

The dependence of $P_N(f, \Delta f)$ on Δf is a consequence of the fact that a small value of Δf implies a long effective integration time. In a long integration time, the output signal can be sampled independently many times, leading to a decrease in the rms noise voltage by the square root of the number of independent measurements. It is found that for many detectors the NEP is also proportional to the square root of the detector area A. This dependence occurs for detectors that produce a background signal (or a dark signal) that is proportional to the detector area, since the fluctuations in the background signal are proportional to the square root of the magnitude of the background signal and thus to $\sqrt{A}$.

For those cases in which the NEP is proportional to the square root of the electrical bandwidth Δf and to the square root of the detector area A, it is possible to introduce a new descriptor that is independent of both Δf and A. This new descriptor, called the specific detectivity, is defined by

$$D^* = \frac{\sqrt{A\,\Delta f}}{P_N}. \tag{7.14}$$

Unlike the NEP, this descriptor is a figure of merit and increases with the sensitivity of the detector.

7.4 ELECTRICAL BANDWIDTH

The definition of the NEP involves the perhaps unfamiliar notion of the electrical bandwidth (sometimes called the noise-equivalent bandwidth) Δf

of the detection system. If the detection system responded uniformly to modulation frequencies in the range f_1 to f_2 and had no response to modulation frequencies outside of this range, it would be natural to take the electrical bandwidth to be $\Delta f = f_2 - f_1$. For the case in which the response varies continuously with f, it is possible to generalize the notion of the electrical bandwidth through use of the definition

$$\Delta f = \int_0^\infty \left| \frac{\mathcal{R}(f)}{\mathcal{R}_{max}} \right|^2 df, \tag{7.15}$$

where $\mathcal{R}_{max}$ is the maximum value of the function $\mathcal{R}(f)$. The electrical bandwidth Δf is defined in terms of the square of $\mathcal{R}(f)$ so as to be a measure of the power in the electrical signal produced by the detection system.

For a system characterized by an exponential decay time τ, the frequency response is given by Eq. (7.5), and the electrical bandwidth is thus given by

$$\Delta f = \int_0^\infty \frac{df}{1 + (2\pi f \tau)^2} = \frac{1}{4\tau}. \tag{7.16}$$

In the following chapters, we shall often be interested in detection systems that integrate the electrical output signal over a sampling time T. The response of such a system to a δ-function input at time $t = -T/2$ is of the form

$$v(t) = \begin{cases} 1 & -T/2 < t < T/2 \\ 0 & \text{otherwise.} \end{cases}$$

The frequency response of the system is thus given by

$$\mathcal{R}(f) = \int_{-T/2}^{T/2} \frac{\mathcal{R}_0}{T} e^{-i2\pi f t} \, dt$$

$$= \mathcal{R}_0 \frac{\sin \pi f T}{\pi f T}. \tag{7.17}$$

The electrical bandwidth Δf is then given using Eq. (7.15) by

$$\Delta f = \int_0^\infty \left(\frac{\sin \pi f T}{\pi f T} \right)^2 df$$

$$= \frac{1}{2T}. \tag{7.18}$$

7.5 NOISE REDUCTION THROUGH AVERAGING

A standard laboratory procedure for making an accurate determination of a signal level in the presence of noise is to average together many independent measurements of the quantity to be determined. Under quite general circumstances, it is found that using this procedure the signal-to-noise ratio of the measurement can be increased by a factor of $n^{1/2}$, where n denotes the number of independent measurements that are averaged together. This section provides a general proof of this result. It should be noted that this result was anticipated earlier in Eq. (7.12), which states that the NEP of a detector decreases as $T^{-1/2}$ as the integration time $T \sim 1/\Delta f$ is increased. This conclusion follows from the fact that the number n of independent measurements increases linearly with the integration time T.

Let x denote a stochastic quantity whose mean value is to be determined. We let the probability distribution for x be denoted $p(x)$ such that $p(x') \, dx'$ denotes the probability that a single measurement of x will yield a value within dx' of x'. Hence the mean of x is given by

$$\bar{x} = \int_{-\infty}^{\infty} x \, p(x) \, dx, \tag{7.19}$$

and the variance of x is given by

$$\overline{(\Delta x)^2} = \int_{-\infty}^{\infty} (x - \bar{x})^2 p(x) \, dx. \tag{7.20}$$

Let us assume that n independent measurements of x are performed and that these results are averaged together to form a quantity y where

$$y = \frac{1}{n} \sum_{j=1}^{n} x_j, \tag{7.21}$$

with x_j denoting the result of the jth measurement. We wish to determine the mean and variance of y. The mean of y is given by

$$\bar{y} = \int dx_1 \int dx_2 \cdots \int dx_n \, y \, p(x_1) p(x_2) \cdots p(x_n).$$

$$= \int dx_1 \int dx_2 \cdots \int dx_n \frac{1}{n} \sum_{j=1}^{n} x_j p(x_1) p(x_2) \cdots p(x_n)$$

$$= \bar{x}, \tag{7.22}$$

since the second-to-last form is simply $1/n$ times the summation of n terms of the form of Eq. (7.19). The variance of y can be calculated similarly and is given by

$$\overline{(\Delta y)^2} = \int dx_1 \int dx_2 \cdots \int dx_n (y - \bar{y})^2 p(x_1) p(x_2) \cdots p(x_n). \quad (7.23)$$

The quantity $(y - \bar{y})^2$ appearing here can be expressed as

$$(y - \bar{y})^2 = \left[\left(\frac{1}{n} \sum_{j=1}^{n} x_j \right) - \bar{y} \right]^2$$

$$= \left[\frac{1}{n} \sum_{j=1}^{n} (x_j - \bar{x}) \right]^2$$

$$= \left[\frac{1}{n^2} \sum_{j=1}^{n} (x_j - \bar{x})^2 + \sum_{j=1}^{n} \sum_{i \neq j} (x_i - \bar{x})(x_j - \bar{x}) \right]. \quad (7.24)$$

In evaluating Eq. (7.22), the second term in expression (7.23) gives a contribution of zero since it gives rise to integrals of the form

$$\int_{-\infty}^{\infty} (x_i - \bar{x}) p(x_i) \, dx_i,$$

which, according to Eq. (7.19), equals zero. The first term in expression (7.23) produces n identical integrals of the form of Eq. (7.20), and thus the variance of y is given by

$$\overline{(\Delta y)^2} = \frac{\overline{(\Delta x)^2}}{n}, \quad (7.25)$$

proving that the variance of the mean decreases in proportion to the number of measurements that are averaged together. It should be noted that this conclusion is independent of the functional form of $p(x)$ as long as the mean and variance of x are finite.

BIBLIOGRAPHY

F. Grum and R. J. Becherer, *Optical Radiation Measurements*, Academic, New York, 1979.

R. D. Hudson, *Infrared System Engineering*, Wiley, New York, 1969.

R. C. Jones, in *Advances in Electronics*, Vol. V, Academic, New York, 1953.

R. J. Keyes, ed., *Optical and Infrared Detectors*, Springer, Berlin, 1977.

R. H. Kingston, *Detection of Optical and Infrared Radiation*, Springer, Berlin, 1978.

P. W. Kruse, L. D. McGlauchlin, and R. B. McQuistan, *Infrared Technology*, Wiley, New York, 1962.

R. A. Smith, F. E. Jones, and R. P. Chasmar, *The Detection and Measurement of Infrared Radiation*, Oxford University, London, 1968.

A. Stimson, *Photometry and Radiometry for Engineers*, Wiley, New York, 1974.

E. B. Wilson, Jr., *An Introduction to Scientific Research*, McGraw-Hill, New York, 1952. Chapters 7–9 discuss the measurement of stochastic quantities.

W. L. Wolfe and G. J. Zissis, eds. *The Infrared Handbook*, Office of Naval Research, Arlington, Va, 1978.

PROBLEMS

1 Consider the detection of a signal of the form

$$P(t) = \begin{cases} 0 & t < 0 \\ P_S e^{-t/\tau_S} & t \geqslant 0 \end{cases}$$

using a detector of response time τ whose NEP, referred to a 1-Hz electrical bandwidth, is given by P_N. Assume that P_N is independent of the modulation frequency f and scales as $(\Delta f)^{1/2}$.

(a) In the absence of noise, what is the output pulse shape?

(b) Discuss qualitatively how noise will affect the resulting output pulse for all values of the ratios τ/τ_S and of P_N/P_S

(c) Derive expressions, in terms of the quantities τ, τ_S, P_N, and P_S, for the following quantities:

   (i) The signal-to-noise ratio of a measurement performed by averaging the output of the detector over the time interval 0 to τ.

   (ii) The signal-to-noise ratio of a measurement performed by averaging the output of the detector over the time interval 0 to τ_S.

2 Comment, on the basis of your results for Problem 1, on how noise limits our ability to perform an accurate determination of P_S and of τ_S.

3 Assume that P_N depends on f as

$$P_N \sim \left(1 + 4\pi^2 f^2 \tau^2\right)^{1/2}.$$

How are the results of Problems 1 and 2 changed?

8

Noise in the Detection Process

This chapter discusses the mathematic description of stochastic processes and illustrates the general principles of stochastic processes by treating two commonly encountered noise sources in detection systems: shot noise (which is closely related to photon noise) and Johnson noise.

8.1 MATHEMATICAL DESCRIPTION OF NOISE

Figure 8.1a shows the voltage signal $v(t)$ that might be produced by a certain radiation detector. One might describe this signal in terms of its time-averaged value $\bar{v}$ and in terms of the rms deviation $(\Delta v)_{rms} = \{\overline{[v(t) - \bar{v}]^2}\}^{1/2}$ of the signal from its average value. The quantity $(\Delta v)_{rms}$ is thus a measure of the noise in the electrical signal. The quantity $\overline{(\Delta v)^2} = [(\Delta v)_{rms}]^2$, known as the variance of $v(t)$, is another common descriptor of the noise in this signal.

In many circumstances, one is interested in more than just the average value and rms deviation of a fluctuating quantity. Figure 8.1b, for instance, shows another voltage signal that has the same values of $\bar{v}$ and of $(\Delta v)_{rms}$ as those shown in Fig. 8.1a, but is qualitatively quite different because the characteristic time scales of the fluctuations differ for the two cases. The noise has different frequency components in the two cases, and it is useful to be able to describe its frequency spectrum in a quantitative manner.

Spectral Densities

The simplest method of specifying the harmonic content of a signal $v(t)$ is through its Fourier transform

$$V(f) = \int_{-\infty}^{\infty} v(t) e^{-i2\pi ft} \, dt. \tag{8.1a}$$

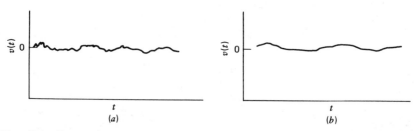

Figure 8.1. Two stochastic voltage signals, both having zero mean and the same rms deviation from zero, but composed of different frequency components.

The signal $v(t)$ can be recovered from $V(f)$ by the inverse transform

$$v(t) = \int_{-\infty}^{\infty} V(f)e^{i2\pi ft}\,df. \tag{8.1b}$$

However, if $v(t)$ is a stationary, stochastic function of time defined for all values of t in the interval $-\infty$ to ∞, $v(t)$ is not square-integrable, and thus its Fourier transform does not exist. The harmonic content of $v(t)$ can, however, be specified by a quantity known either as the spectral density of $v(t)$ or as the power spectrum of $v(t)$. We first define the truncated signal $v_T(t)$ by

$$v_T(t) = \begin{cases} v(t) & \text{for } \dfrac{-T}{2} < t < \dfrac{T}{2} \\ 0 & \text{otherwise.} \end{cases} \tag{8.2a}$$

The Fourier transform $V_T(f)$ of $v_T(t)$ does exist and is given by

$$V_T(f) = \int_{-\infty}^{\infty} v_T(t)e^{-i2\pi ft}\,dt$$

$$= \int_{-T/2}^{T/2} v(t)e^{-i2\pi ft}\,dt. \tag{8.2b}$$

The spectral density $S(f)$ of $v(t)$ is then defined most generally as

$$S(f) = \lim_{T \to \infty} \left\langle \frac{1}{T}|V_T(f)|^2 \right\rangle,$$

where the angular brackets denote an ensemble average. That is, they require that the enclosed quantity be averaged over an ensemble of identically prepared systems. The spectral density must, in general, be defined as an ensemble average in this way to ensure that $S(f)$ converges to a smooth function of f for arbitrary $v(t)$. However, the necessity of performing this ensemble average greatly complicates the calculation of the spectral density. In practice, one is never able to measure the quantity $S(f)$. Instead one measures $S(f)$ averaged over some (perhaps very small) frequency interval. Averaged over an arbitrarily small frequency interval, the quantity

$$S(f) = \lim_{T \to \infty} \frac{1}{T} |V_T(f)|^2 \qquad (8.3)$$

converges for any physically realizable random signal. Thus, for reasons of convenience, we shall use Eq. (8.3) as the definition of the spectral density in the following discussion.

The spectral density $S(f)$ can be interpreted physically as the power per unit frequency interval that a voltage signal $v(t)$ could deliver to a 1 Ω resistor, as shown in Fig. 8.2. This connection is established by noting that the average power delivered by $v(t)$ is given by

$$\overline{v^2} = \lim_{T \to \infty} \frac{1}{T} \int_{T/2}^{T/2} v^2(t)\, dt$$

$$= \lim_{T \to \infty} \frac{1}{T} \int_{-\infty}^{\infty} v_T^2(t)\, dt.$$

As a consequence of Parseval's theorem, which states that

$$\int_{-\infty}^{\infty} |x(t)|^2\, dt = \int_{-\infty}^{\infty} |X(f)|^2\, df \qquad (8.4)$$

for any Fourier transform pair $x(t)$ and $X(f)$, and using the fact that $v_T(t)$ is real, the expression for $\overline{v^2}$ can be expressed as

$$\overline{v^2} = \lim_{T \to \infty} \frac{1}{T} \int_{-\infty}^{\infty} |V_T(f)|^2\, df,$$

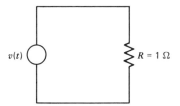

$v(t)$ $R = 1\ \Omega$

Figure 8.2. The power delivered to the resistor per unit frequency interval is given by $S(f)$.

or, by interchanging the limit and the integration and introducing Eq. (8.3), as

$$\overline{v^2} = \int_{-\infty}^{\infty} S(f)\, df. \tag{8.5}$$

Correlation Functions

It is often awkward to carry out the limiting process required by Eq. (8.3), and it is thus fortunate that the spectral density can be related to the more easily calculated correlation function. The correlation function (or, more specifically, the autocorrelation function) is defined most generally as

$$C(t_1, t_2) = \langle v(t_1)v(t_2)\rangle,$$

where, as before, the angular brackets denote an ensemble average. In the study of electronic noise in detection systems, we shall most often be interested in random signals that are ergodic (i.e., random signals for which an ensemble average is equivalent to a time average). A signal that is ergodic must also be stationary in the sense that $C(t_1, t_2)$ can depend on the times only as the difference $\tau = t_2 - t_1$. Thus the correlation function of an ergodic random signal can be expressed as

$$C(\tau) = \lim_{T \to \infty} C_T(\tau), \tag{8.6a}$$

where

$$C_T(\tau) = \frac{1}{T}\int_{-T/2}^{T/2} v(t)v(t + \tau)\, dt. \tag{8.6b}$$

The correlation function $C(\tau)$ can be represented symbolically by

$$C(\tau) = \overline{v(t)v(t + \tau)}, \tag{8.7}$$

where the bar denotes a time average. Thus, evaluated at $\tau = 0$, it has the value

$$C(0) = \overline{v^2}. \tag{8.8}$$

The correlation function is an even function of τ for any stationary process, since

$$C(-\tau) = \overline{v(t)v(t - \tau)}$$

$$= \overline{v(t + \tau)v(t)} = C(\tau), \tag{8.9}$$

where the second equality results from a shift in the time axis. The correlation function also has the property

$$C(0) \geq |C(\tau)|. \tag{8.10}$$

This can be demonstrated using the inequality

$$[v(t) \pm v(t + \tau)]^2 \geq 0$$

which, when expanded, becomes

$$v^2(t) + v^2(t + \tau) \geq \pm 2v(t)v(t + \tau).$$

By taking a time average of each side, we obtain

$$\overline{v^2} \geq \pm \overline{v(t)v(t + \tau)},$$

from which Eq. (8.10) follows.

Wiener–Khintchine Theorem

The Wiener–Khintchine theorem states that the quantities $S(f)$ and $C(\tau)$ form a Fourier transform pair. This theorem is demonstrated by first calculating the transform of $C_T(\tau)$:

$$\int_{-\infty}^{\infty} C_T(\tau) e^{-i2\pi f\tau} d\tau = \frac{1}{T} \int_{-\infty}^{\infty} d\tau \int_{-\infty}^{\infty} dt \, v_T(t) v_T(t + \tau) e^{-i2\pi f\tau}$$

$$= \frac{1}{T} \int_{-\infty}^{\infty} dt \, v_T(t) e^{i2\pi ft} \int_{-\infty}^{\infty} d\tau \, v_T(t + \tau) e^{-i2\pi f(t+\tau)}$$

$$= \frac{1}{T} V_T(-f) V_T(f),$$

or

$$\int_{-\infty}^{\infty} C_T(\tau) e^{-i2\pi f\tau} d\tau = \frac{1}{T} |V_T(f)|^2. \tag{8.11}$$

This result demonstrates a special case of the convolution theorem, which states that the Fourier transform of the convolution of two functions equals the product of the Fourier transforms of the two functions. By taking the limit of Eq. (8.11) as $T \to \infty$ and using Eqs. (8.3) and (8.6), we obtain

$$S(f) = \int_{-\infty}^{\infty} C(\tau)e^{-i2\pi f\tau}\,d\tau, \qquad (8.12)$$

and by taking the inverse transform of $S(f)$, we obtain

$$C(\tau) = \int_{-\infty}^{\infty} S(f)e^{i2\pi f\tau}\,df. \qquad (8.13)$$

Equations (8.12) and (8.13) constitute the Wiener–Khintchine relations. Since, by Eq. (8.9), $C(\tau)$ is an even function of τ, Eq. (8.12) can be rewritten as

$$S(f) = 2\int_{0}^{\infty} C(\tau)\cos 2\pi f\tau\,d\tau, \qquad (8.14)$$

showing that $S(f)$ is an even function of f and allowing Eq. (8.13) to be rewritten as

$$C(\tau) = 2\int_{0}^{\infty} S(f)\cos 2\pi f\tau\,df. \qquad (8.15)$$

Equations (8.14) and (8.15) constitute another form of the Wiener–Khintchine relations.

In practical applications, it is often convenient to work with frequencies f defined for the interval 0 to ∞ instead of the interval $-\infty$ to ∞. The spectral density of $v(t)$ is often expressed then as a function of the frequency f through use of the obvious symbolic notation

$$\overline{v^2}(f) = 2S(f). \qquad (8.16)$$

Using this notation, the Wiener–Khintchine relations become

$$\overline{v^2}(f) = 4\int_{0}^{\infty} C(\tau)\cos 2\pi f\tau\,d\tau, \qquad (8.17)$$

$$C(\tau) = \int_{0}^{\infty} \overline{v^2}(f)\cos 2\pi f\tau\,df. \qquad (8.18)$$

and Eq. (8.5) for the mean square of v becomes

$$\overline{v^2} = \int_{0}^{\infty} \overline{v^2}(f)\,df.$$

Identical Events Occurring at Random Times

As an illustration of the general principles described earlier, we consider here the important example of a signal $v(t)$ composed of a very large number of identical but independent events, each of which occurs at a random instant of time. If the random event is the emission of an electron from a forward-biased, temperature-limited thermionic diode, this analysis can be used to describe a common noise source known as shot noise or, by a curious historical accident, as Schottky noise, after its discoverer. If the random event is the scattering of an electron in a resistor, this analysis can be used to describe Johnson noise; if the random event is the absorption of a photon from a beam of constant power, this analysis can be used to describe photon noise in the detection process.

We shall assume that the detection system is linear, so that if the output signal due to a single event occurring at time $t = 0$ is given by $g(t)$, as shown in Fig. 8.3, the output due to N such events occurring in the time interval $-T/2 \leqslant t \leqslant T/2$ is given by

$$v_T(t) = \begin{cases} \displaystyle\sum_{i=1}^{N} g(t - t_i) & \dfrac{-T}{2} \leqslant t \leqslant \dfrac{T}{2} \\ 0 & \text{otherwise.} \end{cases} \tag{8.19}$$

Here N is a random variable giving the number of such events in the time interval, and t_i is a random variable giving the time at which the ith event

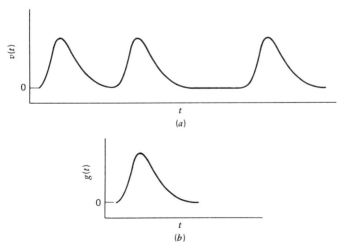

Figure 8.3. Signal $v(t)$ is composed of identical pulses $g(t)$ occurring at random instants of time.

occurred. The time average of $v(t)$ is then given by

$$\bar{v} = \lim_{T \to \infty} \frac{1}{T} \int_{-T/2}^{T/2} \sum_{i=1}^{N} g(t - t_i) \, dt, \qquad (8.20)$$

which can be expressed as

$$\bar{v} = r \int_{-\infty}^{\infty} g(t) \, dt, \qquad (8.21)$$

where r is the average rate at which the events occur and is given by

$$r = \left\langle \frac{N}{T} \right\rangle$$

or as

$$r = \lim_{T \to \infty} \frac{N}{T}, \qquad (8.22)$$

since the system has been assumed to be ergodic.

The statistical properties of $v(t)$ can be expressed in terms of the correlation function of $v(t)$ given by Eq. (8.6) as

$$C_v(\tau) = \lim_{T \to \infty} \frac{1}{T} \int_{-T/2}^{T/2} v(t) v(t + \tau) \, dt$$

$$= \lim_{T \to \infty} \frac{1}{T} \int_{-\infty}^{\infty} \sum_{i=1}^{N} g(t - t_i) \sum_{j=1}^{N} g(t + \tau - t_j) \, dt$$

$$= \lim_{T \to \infty} \frac{1}{T} \int_{-\infty}^{\infty} \left[\sum_{i=1}^{N} g(t - t_i) g(t + \tau - t_i) \right.$$

$$\left. + \sum_{\substack{i=1 \\ }}^{N} \sum_{\substack{j=1 \\ j \neq i}}^{N} g(t - t_i) g(t + \tau - t_j) \right] dt.$$

The first term in the expression in square brackets gives rise to N identical integrals that may be interpreted as the correlation of $g(t)$ with itself, while the second term may be interpreted as $\bar{v}^2$, since no correlation exists between independent pulses. The correlation function of $v(t)$ thus becomes

$$C_v(\tau) = r \int_{-\infty}^{\infty} g(t) g(t + \tau) \, dt + \bar{v}^2. \qquad (8.23)$$

One measure of the noise in the signal $v(t)$ is its variance

$$\overline{(\Delta v)^2} \equiv \overline{[v(t) - \bar{v}]^2} = \overline{v^2(t)} - \bar{v}^2. \tag{8.24}$$

Since by Eq. (8.8), $\overline{v^2(t)} = C_v(0)$, the variance is given by

$$\overline{(\Delta v)^2} = r \int_{-\infty}^{\infty} g^2(t)\, dt. \tag{8.25}$$

This result, along with Eq. (8.21) for $\bar{v}$, is known as Campbell's theorem.

The spectral density of the signal $v(t)$ can be obtained by introducing the form (8.23) of $C_v(\tau)$ into Eq. (8.12) to obtain

$$S(f) = \int_{-\infty}^{\infty} \left[\left(r \int_{-\infty}^{\infty} g(t)g(t + \tau)\, dt \right) + \bar{v}^2 \right] e^{-i2\pi f\tau}\, d\tau. \tag{8.26}$$

The double integral in this expression can be evaluated using a method identical to that used to obtain Eq. (8.11), (or using the convolution theorem of Fourier analysis), while the contribution from $\bar{v}^2$ can be obtained from the standard definition of the δ function as

$$\int_{-\infty}^{\infty} e^{-i2\pi f\tau}\, d\tau = \delta(f), \tag{8.27a}$$

where

$$\int_{-\infty}^{\infty} \delta(f) e^{i2\pi f\tau}\, df = 1, \tag{8.27b}$$

giving the result

$$S(f) = r|G(f)|^2 + \bar{v}^2 \delta(f), \tag{8.28}$$

where the Fourier transform of $g(t)$ has been designated

$$G(f) = \int_{-\infty}^{\infty} g(t) e^{-i2\pi ft}\, dt. \tag{8.29}$$

This result is known as Carson's theorem (J. R. Carson, *Bell Syst. Tech. J.*, **10**, 374 (1931)] The term in Eq. (8.28) involving $\delta(f)$ gives the average power that is contained in the dc component of $v(t)$. The other term, which can be considered a representation of the noise in $v(t)$, is related to the spectral content of each random event.

8.2 PHOTON NOISE

Let us for the present consider only the noise that is introduced into any photon detection system as a result of the discrete nature of the radiation field. This noise is fundamental in the sense that it arises not from any imperfection in the detector or its associated electronics but rather from the detection process itself.

This noise process can be understood from two conceptually distinct but complementary points of view. The noise can be viewed as resulting either from the randomness in the arrival times of individual photons or from the randomness in the emission time of photoelectrons. The former point of view forces us to treat the radiation field as a quantized system and thus to consider certain aspects of the detection process (e.g., those related to the Bose–Einstein nature of the radiation field) that are irrelevant for our present purposes. (These aspects are treated in the more general discussion of Chapter 14.) Therefore, for the present, we shall consider the noise as resulting from the randomness of the emission times of the photoelectrons. It will be shown in Chapter 14 that the present analysis holds rigorously when the power fluctuations in the incident radiation can be ignored, and that this will be true for thermal radiation whenever the mean occupation number of the radiation is much less than unity. According to Eqs. (3.57) and (3.62), this will occur when the temperature of the field is related to the frequency of the detected radiation by $h\nu \gg kT$.

We here consider the case of nearly monochromatic radiation of frequency ν and of constant power P falling onto a photon detector. The detector is assumed to be an *ideal photon detector* in the sense that it produces no output current in the absence of incident power and no noise except that related to the randomness of the emission times of the photoelectrons. We shall assume that these emission events are probabilistic in the sense that they are uncorrelated and occur at the average rate

$$r = \frac{\eta P}{h\nu}, \tag{8.30}$$

where η is the detector quantum efficiency. The average number of photo-events occurring in the time T is thus

$$\overline{N} = rT, \tag{8.31}$$

but, due to the probabilistic nature of the emission process, the actual number emitted in any one particular interval of length T will fluctuate around this value. The probability $p(n)$ that in any one such interval exactly

n photoevents occur is given by the Poisson probability distribution. This probability distribution can be derived by dividing the time interval T into some large number n of identical segments. The average number of photoevents occurring in any one segment is thus equal to $\overline{N}/n$. For n sufficiently large, this number is much less than unity, and for this case $\overline{N}/n$ can be interpreted as the probability that one event occurs in a given segment. The probability that exactly N photoevents occur in the total interval T is then given by the binomial distribution*

$$p_n(N) = \frac{n!}{N!(n-N)!}\left(\frac{\overline{N}}{n}\right)^N\left(1 - \frac{\overline{N}}{n}\right)^{n-N}.$$

In this expression, the factor $(\overline{N}/n)^N$ represents the probability that photoevents occur in N specific segments, and the factor $(1 - \overline{N}/n)^{n-N}$ represents the probability that events do not occur in the remaining $n - N$ segments. The factor $n!/N!(n-N)!$ denotes a combination of n things taken N at a time and represents the total number of ways in which N indistinguishable events can occur in n segments. This expression can be simplified somewhat by dividing the first factor and multiplying the second factor by n^N, giving

$$p_n(N) = \frac{1\left(1 - \frac{1}{n}\right)\left(1 - \frac{2}{n}\right)\cdots\left(1 - \frac{N+1}{n}\right)}{N!}\overline{N}^N\left(1 - \frac{\overline{N}}{n}\right)^{n-N}$$

Finally, the Poisson distribution is obtained by taking the limit of $p_n(N)$ as $n \to \infty$:

$$p(N) = \lim_{n\to\infty} p_n(N)$$

$$= \frac{\overline{N}^N}{N!}\lim_{n\to\infty}\left(1 - \frac{\overline{N}}{n}\right)^{n-N}$$

or

$$p(N) = \frac{\overline{N}^N}{N!}e^{-\overline{N}} \tag{8.32}$$

This probability distribution is illustrated in Fig. 8.4.

---

*This distribution is discussed in most books on probability theory. See, e.g., the book by Papoulis listed in the bibliography for this chapter.

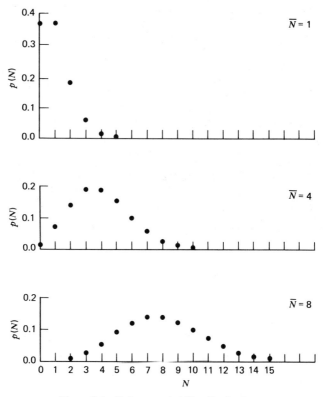

Figure 8.4. Poisson probability distribution.

It can readily be verified that the Poisson distribution is properly normalized since

$$\sum_{N=0}^{\infty} p(N) = e^{-\overline{N}} \sum_{N=0}^{\infty} \frac{\overline{N}^{N}}{N!} = e^{-\overline{N}} e^{\overline{N}} = 1. \qquad (8.33)$$

In addition, the parameter $\overline{N}$ of (8.32) is in fact the expectation value of N, since

$$\sum_{N=0}^{\infty} Np(N) = \sum_{N=1}^{\infty} N \frac{\overline{N}^{N}}{N!} e^{-\overline{N}} = \sum_{N=1}^{\infty} \frac{\overline{N}^{N}}{(N-1)!} e^{-\overline{N}}$$

$$= \overline{N} \sum_{N=1}^{\infty} \frac{\overline{N}^{N-1}}{(N-1)!} e^{-\overline{N}}$$

$$= \overline{N} \sum_{M=0}^{\infty} \frac{\overline{N}^{M}}{M!} e^{-\overline{N}} = \overline{N}. \qquad (8.34)$$

The second to last form of this expression followed from the substitution $M = N - 1$. The Poisson distribution often contains more details of the detection statistics than are necessary; it is often sufficient to specify only $\overline{N}$ and the variance of N, given by

$$\overline{(\Delta N)^2} = \overline{(N - \overline{N})^2} = \overline{N^2} - \overline{N}^2. \tag{8.35}$$

The quantity $\overline{N^2}$ is calculated, using the identity $N^2 = N + N(N - 1)$, as

$$\overline{N^2} = \sum_{N=0}^{\infty} Np(N) + \sum_{N=0}^{\infty} N(N - 1)p(N). \tag{8.36}$$

Here the first summation is, by Eq. (8.34), equal to $\overline{N}$, and the second summation can be evaluated to give

$$\sum_{N=0}^{\infty} N(N - 1)p(N) = \sum_{N=2}^{\infty} N(N - 1)\frac{\overline{N}^N e^{-\overline{N}}}{N!}$$

$$= \overline{N}^2 \sum_{N=2}^{\infty} \frac{\overline{N}^{N-2} e^{-\overline{N}}}{(N - 2)!}$$

$$= \overline{N}^2 \sum_{M=0}^{\infty} \frac{\overline{N}^M e^{-\overline{N}}}{M!} = \overline{N}^2, \tag{8.37}$$

where $M = N - 2$ has been introduced in the second to last form. The variance is thus given by

$$\overline{(\Delta N)^2} = \overline{N}, \tag{8.38}$$

and the rms fluctuation is given as $(\Delta N)_{\text{rms}} = [\overline{(\Delta N)^2}]^{1/2} = \overline{N}^{1/2}$. Thus the fractional fluctuation $(\Delta N)_{\text{rms}}/\overline{N}$ of the number of photoevents occurring in a fixed time interval must be equal to $\overline{N}^{-1/2}$. Since $\overline{N} = rT$, the fractional fluctuation varies with the integration time as $T^{-1/2}$ and can be made arbitrarily small by increasing the length of the integration.

The fluctuation in the rate of occurrence of photoevents gives rise to noise in the photocurrent produced by a photon detector. Suppose a photon detection system is characterized by an averaging time T. This means that the current leaving the detector at any instant of time is proportional to the number N of photoevents that have occurred in the preceding time interval T. The average value of the photocurrent is thus

$$\bar{i} = \frac{e\overline{N}}{T}. \tag{8.39}$$

A measure of the noise in the photocurrent is the variance of $i(t)$ or

$$\left(i_N\right)^2 \equiv \overline{\left(\Delta i\right)^2} \equiv \overline{\left(i - \bar{i}\right)^2}, \tag{8.40}$$

which can be evaluated, using Eq. (8.38), as

$$\overline{\left(\Delta i\right)^2} = \frac{e^2}{T^2} \overline{\left(N - \overline{N}\right)^2} = \frac{e^2 \overline{N}}{T^2},$$

or

$$\overline{\left(\Delta i\right)^2} = \frac{e\bar{i}}{T}. \tag{8.41}$$

This result is often expressed in terms of the electrical bandwidth of the detection system, which by Eq. (7.18) is given as $\Delta f = 1/2T$, giving Schottky's formula

$$\overline{\left(\Delta i\right)^2} = 2e\bar{i}\,\Delta f \tag{8.42}$$

[W. Schottky, *Ann. Phys.* (*Leipzig*) **57** 541 (1918)]. This result reflects the fact that the photocurrent is quantized in units of the electron charge e. Photon noise in the limit considered here (i.e., a situation with no fluctuations in the incident power) is sometimes referred to as shot noise, because Eq. (8.42) gives the familiar result for the shot noise in the current passing through a temperature-limited vacuum diode.

Spectral Density of Photon Noise

The analysis just completed assumed that individual current pulses were temporarily unresolved and thus that the number of such events occurring in a given time interval of duration T could be determined unambiguously. Under these conditions, Eq. (8.42) shows that the mean-square current fluctuations scale as the electrical bandwidth Δf; such noise is known as white noise since it is characterized by a constant noise power per unit frequency interval.

A more detailed analysis of photon noise taking the current pulse shape into account can be given using the method outlined in Section 8.1. Using the notation introduced by Eq. (8.16), in which the frequency f runs from zero to infinity, the spectral density of the current noise is given by Eq. (8.28) as

$$\overline{i_N^2}(f) = 2r|G(f)|^2, \tag{8.43}$$

where

$$G(f) = \int_{-\infty}^{\infty} g(t)e^{-i2\pi ft}\, dt. \tag{8.44}$$

We have dropped the term involving $\delta(f)$ from (8.28), since it is related to the average current and not to the noise.

If we assume first that the current pulses are unresolvable and thus have the form

$$g(t) = e\delta(t), \tag{8.45}$$

whose Fourier transform is given by Eq. (8.27b) as

$$G(f) = e, \tag{8.46}$$

the current noise spectral density is given by Eq. (8.43) as

$$\overline{i_N^2}(f) = 2re^2 = 2e\bar{i}, \tag{8.47}$$

and thus the total noise power in an electrical bandwidth Δf is given by

$$\overline{(\Delta i)^2} = \overline{i_N^2}(f)\,\Delta f = 2e\bar{i}\,\Delta f, \tag{8.48}$$

in agreement with Schottky's equation (8.42).

Consider now the case in which the current pulse has the form of a decaying exponential:

$$g(t) = \begin{cases} 0 & t < 0 \\ \dfrac{e}{\tau}e^{-t/\tau} & t \geqslant 0, \end{cases} \tag{8.49}$$

as would be the case, for instance, if the output pulse of the detector were passed through a low-pass filter of time constant τ. Equation (7.5) thus gives the Fourier transform of this function as

$$G(f) = \frac{e}{1 + i2\pi f\tau}, \tag{8.50}$$

and the spectral density of the current noise thus has the form

$$\overline{i_N^2}(f) = \frac{2e\bar{i}}{1 + (2\pi f\tau)^2}, \tag{8.51}$$

which is shown in Fig. 8.5. For $f \ll 1/2\pi\tau$, this noise approximates white noise.

Limitations on Detectability Set by Photon Noise

Photon noise sets a limitation on our ability to detect weak signals. Consider the case of a signal power P_S falling onto an ideal photodetector of quantum efficiency η. The average signal current leaving the detector is thus

$$i_S = \frac{\eta e P_S}{h\nu},\tag{8.52}$$

and, by Eq. (8.42), the rms noise in the photocurrent is given by

$$i_N = \sqrt{2ei_S\Delta f}$$

$$= \sqrt{\frac{2\eta e^2 P_S \Delta f}{h\nu}}.\tag{8.53}$$

The signal-to-noise ratio in the photocurrent is then given by

$$\frac{S}{N} \equiv \frac{i_S}{i_N} = \sqrt{\frac{\eta P_S}{2h\nu\,\Delta f}}.\tag{8.54}$$

Since the noise equivalent power (NEP) is defined as the value of P_S that gives a signal-to-noise ratio of unity, the NEP is given as

$$P_N = \frac{2h\nu\,\Delta f}{\eta}.\tag{8.55}$$

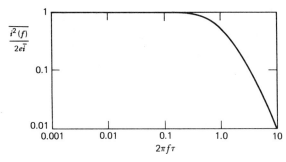

Figure 8.5. Spectral density $\overline{i^2}(f)$ of shot noise normalized by its maximum value $2e\bar{i}$ is shown as a function of f for a system characterized by a single decay time τ.

[The present discussion assumes that constant, i.e., unmodulated power is incident on the detector. In the full notation of Eq. (7.11), the NEP calculated here would be denoted $P_N(0, f)$.] This result can be made intuitive by replacing Δf with $1/2T$, as derived in Eq. (7.18) for an averaging detection system, giving

$$P_N = \frac{h\nu}{\eta T}, \tag{8.56}$$

implying that the minimum detectable power, or NEP, will produce on the average one photon-detection event per measurement time T.

In many situations, the signal to be detected is partially masked by a radiation background that is also incident on the detector. The calculation of the NEP is modified in this circumstance. If signal power P_S and background power P_B fall on a photodetector, the signal current is still given by

$$i_S = \frac{\eta e P_S}{h\nu}, \tag{8.57}$$

but the rms current noise is now given by

$$i_N = \sqrt{\frac{2\eta e^2 (P_S + P_B)\,\Delta f}{h\nu}}, \tag{8.58}$$

and the signal-to-noise ratio is given by

$$\frac{S}{N} = \frac{i_S}{i_N} = \sqrt{\frac{\eta P_S^2}{2h\nu\,\Delta f(P_S + P_B)}}. \tag{8.59}$$

For the limit in which the background power P_B is much greater than the signal power P_S, the NEP can again be determined by setting S/N equal to unity, giving

$$P_N = \sqrt{\frac{2h\nu P_B\,\Delta f}{\eta}}. \tag{8.60}$$

In many circumstances the background power P_B striking the detector is proportional to the area A of the detector; this will be true, for instance, if the background is due to an extended object of sufficient size. For this case, we can characterize the background by its irradiance at the detector as

$$E_B = \frac{P_B}{A},$$

(8.61)

and the specific detectivity, defined in Eq. (7.14), is given by

$$D^* = \frac{\sqrt{A \, \Delta f}}{P_N} = \sqrt{\frac{\eta}{2 h \nu E_B}}.$$

(8.62)

Equations (8.55) and (8.60) give the NEP of an ideal photon detector (i.e., one limited only by photon noise) under two different operating conditions. An actual photodetector may be worse than ideal in the sense that it produces noise by means of additional mechanisms. However, if the background power is sufficiently large that photon noise is the predominant noise source, the NEP is given by Eq. (8.60) and the detector is said to be background-noise limited. Likewise, a detector so good that its NEP is given by Eq. (8.55) is said to be signal-noise limited. It should be noted that in this limiting case the NEP is proportional to Δf, and not to $(\Delta f)^{1/2}$, as in the more common case in which the noise is independent of the signal level.

8.3 JOHNSON NOISE

It has been found experimentally [J. B. Johnson, *Phys. Rev.*, **32**, 110 (1928)] that the thermal agitation of electrons contained in a resistor gives rise to a fluctuating voltage across the leads of the resistor. These fluctuations are known as Johnson noise. In a sense, Johnson noise is less fundamental than photon noise, since it is not an inherent aspect of the detection process; it is often of practical importance, however, in limiting the sensitivity of detection systems. Many photodetectors can be considered to be current sources, and, as shown in Fig. 8.6, a voltage output is

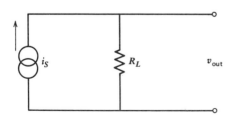

Figure 8.6. Signal current, load resistor, and output voltage are illustrated.

produced by passing the detector current i_S through a load resistor R_L. The Johnson noise produced by the load resistor will thus be superposed on the signal voltage $i_S R_L$.

Johnson noise can be analyzed through consideration of the simple circuit shown in Fig. 8.7, in which all the components are in equilibrium at temperature T. This circuit can be thought of as a classical system with one degree of freedom: the voltage across the capacitor. The equipartition theorem then requires that the average energy stored in that degree of freedom, $\frac{1}{2}C\overline{v_N^2}$, be equal to $\frac{1}{2}kT$ and thus that

$$\overline{v_N^2} = \frac{kT}{C}. \tag{8.63}$$

Thus a resistor at a temperature of 300 K that is shunted by a capacitor of value 10 pF, will produce an rms noise voltage given by the square root of this expression as 20 μV.

The spectral density of this voltage can be obtained by noting that since the transient behavior of the circuit is characterized by an exponential decay time given by the product RC, the correlation function of the voltage fluctuations must be given by

$$C(\tau) = \overline{v(t)v(t+\tau)}$$

$$= \overline{v(t)v(t)}\,e^{-|\tau|/RC}$$

$$= \overline{v_N^2}\,e^{-|\tau|/RC}. \tag{8.64}$$

[This same conclusion can be reached in a more rigorous fashion by considering each electron collision within the resistor to produce an output pulse with the form of a decaying exponential and using Eqs. (8.23) and (8.25) for the correlation function of a collection of identical events occur-

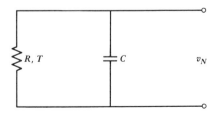

Figure 8.7. Calculation of Johnson voltage noise.

ring at random instants of time.] The spectral density of the noise voltage is then given by Eq. (8.17) as

$$\overline{v_N^2}(f) = 4\int_0^\infty \overline{v_N^2}\, e^{-\tau/RC} \cos 2\pi f \tau\, d\tau$$

$$= \frac{4\,\overline{v_N^2}\,RC}{1 + (2\pi fRC)^2}$$

$$\simeq 4\,\overline{v_N^2}\,RC, \tag{8.65}$$

where the limit $f \ll 1/RC$ has been assumed in the last line. Using Eq. (8.63), this result becomes

$$\overline{v_N^2}(f) = 4kTR, \tag{8.66}$$

which is known as Nyquist's formula. Since the value of C does not appear here, this formula must hold in the same form in the limit $C \to 0$, in which case the requirement $f \ll 1/RC$ is satisfied for all values of the frequency. Johnson noise is thus another example of white noise, since its spectral density is independent of f. [This conclusion is valid only for frequencies such that $hf \ll kT$. The general result is given in Eq. (8.71)].

Figure 8.8 shows two different equivalent-circuit representations of a real resistor whose temperature is T. The physical resistor can be modeled as a noiseless resistor either in series with a noise voltage source or in parallel with a noise current source.

An alternative treatment of Johnson noise, due originally to H. Nyquist [*Phys. Rev.*, **32**, 110 (1928)], considers Johnson noise to be a one-dimensional form of blackbody radiation. Figure 8.9 shows a lossless transmission line of characteristic impedance Z, terminated at each end by resistors of value $R = Z$ that are maintained at equal temperatures T. (One might think of this as a 50-Ω coaxial cable, where at each end the inner and outer conductors are connected through 50-Ω resistors). Each resistor will produce a thermal noise voltage that causes a noise power $P_f\, df$ in the frequency interval df to be transmitted through the line to the resistor at the other end. Since the line is terminated at both ends, there are no reflection losses, and all the power carried by the line is dissipated at the terminating resistors.

This amount of power $P_f\, df$ can be calculated by considering this power to be blackbody radiation confined to a one-dimensional cavity. The calculation is analogous to that given in Chapter 3 for the usual three-

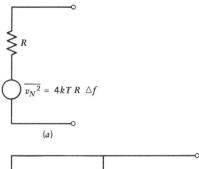

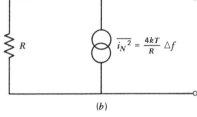

Figure 8.8. Equivalent-circuit representations of a physical resistor in terms of a noiseless resistor and (a) a noise voltage source and (b) a noise current source.

dimensional case. We first need to calculate the number of modes of the electromagnetic field per unit length of the transmission line. This number cannot depend on the details of how the transmission line is terminated, and the calculation is simplified if it is assumed that the two conductors are shorted together at both ends of the transmission line so that it can support standing waves of the form

$$v(z, t) = v_0 \sin kz \sin 2\pi ft, \qquad (8.67)$$

where

$$k = \frac{2\pi f}{c/N}. \qquad (8.68)$$

Figure 8.9. Transmission-line treatment of Johnson noise.

The vanishing of the voltage at both ends further requires that

$$k = \frac{m\pi}{L},\tag{8.69}$$

where m is any integer and L is the length of the transmission line. Since the allowed values of k are thus spaced by π/L, the number of allowed modes with wavevectors between k and $k + dk$ is equal to $dk\, L/\pi$, and the number of allowed modes with frequency between f and $f + df$ is thus equal to $2df\, L/(c/N)$. The density of modes per unit length and per unit frequency interval is then given by

$$\rho_{1-d}(f) = \frac{2}{c/N},\tag{8.70}$$

and the energy per unit length and per unit frequency interval is given by the product of Eq. (8.70) with the average energy per mode:

$$u_{1-d}(f) = \frac{2}{c/N}\frac{hf}{e^{hf/kT} - 1}.\tag{8.71}$$

The power flow in either direction is one-half this result multiplied by the velocity c/N of propagation, giving

$$P_f = \frac{hf}{e^{hf/kT} - 1}$$

$$\approx kT \quad \text{for } hf \ll kT.\tag{8.72}$$

The spectral density of the noise voltage that would result in such a power

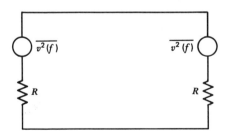

Figure 8.10. Noise-equivalent circuit for the transmission line.

transfer can be determined from Fig. 8.10 as

$$P_f \, df = \frac{\overline{v^2}(df) \, df}{4R}$$

or

$$\overline{v^2}(f) = 4kTR, \tag{8.73}$$

which is in agreement with the result obtained earlier in Eq. (8.66).

BIBLIOGRAPHY

Many of the works cited at the end of Chapter 7 contain a discussion of noise. In addition, the author has found the following list particularly useful.

W. B. Davenport, Jr., and W. L. Root, *Random Signals and Noise*, McGraw-Hill, New York, 1958.

A. Papoulis, *Probability, Random Variables, and Stochastic Processes*, McGraw-Hill, New York, 1965.

F. Reif, *Fundamentals of Statistical and Thermal Physics*, McGraw-Hill, New York, 1965 (see especially Chapter 15).

S. O. Rice, "Mathematical Analysis of Random Noise," *Bell System Tech. J.*, **23**, 282(1944) continued in **24**, 46 (1945), reprinted in *Noise and Stochastic Processes*, N. Wax, ed., Dover, New York, 1954, pp. 133–294.

V. V. Solodovnikov, *Introduction to the Statistical Dynamics of Automatic Control Systems*, Dover, New York, Chapter 3.

PROBLEMS

1 Calculate the correlation function and spectral density for the following cases:

(a) $v(t) = v_0 + v_1 \cos 2\pi f t$,

(b) $v(t) = v_0 \cos(2\pi f t + \phi)$,

where the phase ϕ changes randomly each time the oscillator makes a "phase-changing collision." Assume that the probability that the oscillator has not made such a collision in the time interval T is given by

$$p(T) = e^{-\gamma T}.$$

2 Consider the voltage $v(t)$ appearing across a capacitor $C = 0.01$ μF in parallel with a resistor $R = 1$ MΩ at 300 K as shown in the circuit diagram below. A typical measurement of $v(t)$ might look like the following graph.

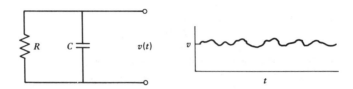

(a) Trace the graph of $v(t)$ onto your solution paper and label the time axis in seconds and the vertical axis in volts as accurately as possible.

(b) C is now replaced by a 0.1-μF capacitor. Sketch a typical trace of $v(t)$ on the same scale as part (a).

(c) C is now replaced with the original capacitor and R is replaced with a 10-MΩ resistor. Sketch a typical trace on the same scale as part (a).

3 Determine the maximum electrical bandwidth (which is approximately the maximum data transfer rate) that can be utilized with an optical communication system composed of a temporally modulated, 10-W peak power laser placed on the moon and a receiver placed on the earth. The receiver consists of a 1-m diameter telescope and an ideal photon detector. The laser wavelength is 0.53 μm, and it is collimated by a 1-m-diameter telescope. It is required that the detection signal-to-noise ratio be 10 : 1.

4 Determine the spectral density $\overline{v_N^2}(f)$ of the Johnson noise produced by a parallel combination of two resistors held at different temperatures, as shown in the following diagram.

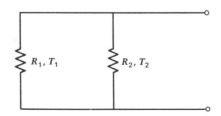

9

Photoemissive Detectors

This chapter discusses that class of detectors which detects radiation by the external photoelectric effect (i.e., the emission of electrons from the surface of a material through the absorption of incident photons). Examples of photoemissive detectors include vacuum photodiodes and photomultipliers, and these detectors are discussed in some detail in this chapter.

9.1 PHOTOELECTRIC EFFECT

It is found that if a beam of monochromatic radiation of sufficiently large frequency ν is allowed to fall onto a metal surface, electrons are emitted from the surface, and the kinetic energy of these electrons is given by

$$K = h\nu - \phi. \tag{9.1}$$

Since the incident photons have energy $h\nu$, the quantity ϕ, known as the work function of the metal, can be interpreted as the binding energy of the electron to the material.

The electronic energy-level structure of a metal is illustrated in Fig. 9.1. Since electrons obey Fermi–Dirac statistics, no more than one electron can occupy a given quantum mechanical state of the system. At the temperature of absolute zero, therefore, all the single-electron states whose energies range from the minimum allowed value (which we shall take as the zero of the energy scale) to a value known as the Fermi energy will be occupied. At room temperature, some electrons will be thermally excited to energies greater than the Fermi energy, but this effect is unimportant for the qualitative discussion presented here. Also shown in the figure is the vacuum level, that is, the energy of an electron located infinitely far from the surface and having zero kinetic energy. The minimum energy required to remove an electron from the metal is thus the difference between the vacuum and Fermi levels and is designated as the work function ϕ. The

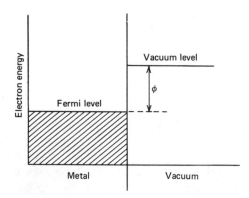

Figure 9.1. Electronic energy-level structure of a metal.

work function of metals is typically several electron volts. The value of 2.0 eV for Cs is the lowest of any elemental metal, and corresponds to the energy of a photon of wavelength 6200 Å.

The photoelectric effect also takes place for semiconductor materials, and in fact all of the common photomultipliers utilize semiconductor photocathodes. The allowed electronic energy levels in a semiconductor are in the form of bands, as shown in Fig. 9.2. Those bands that are entirely filled at absolute zero are called valence bands; those bands that are empty or partly filled are called conduction bands. The energy separating the top of the highest valence band from the bottom of the lowest conduction band is called the energy gap E_g. A Fermi level also exists in a semiconductor. It is defined as the energy at which the probability of occupancy given by the Fermi–Dirac distribution equals $\frac{1}{2}$. The Fermi level usually lies somewhere within the forbidden band. (For degenerate semiconductors, i.e., semicon-

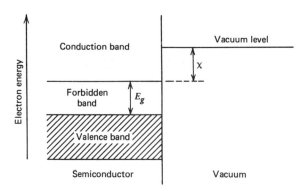

Figure 9.2. Electronic energy-level structure of a semiconductor.

ductors that are very heavily doped, the Fermi level can lie outside of the forbidden band.) The energy separating the bottom of the conduction band from the vacuum level is called the electron affinity χ. Thus the smallest photon energy that can cause the ejection of an electron is given by

$$h\nu = E_g + \chi. \tag{9.2}$$

Photocathodes with enhanced response at wavelengths in the range 0.8–1.5 μm have been developed recently by producing surfaces with a negative electron affinity. By appropriate doping and surface preparation (the details are discussed in the works by Engstrom and Zwicker cited at the end of the chapter), the energy bands are made to bend in the neighborhood of the surface, as shown in Fig. 9.3. An electron excited to the conduction band in the bulk of the material will have an energy greater than that of the vacuum level. Thus if it diffuses to the surface, it will be ejected spontaneously without having to overcome any surface potential. Thus the smallest photon energy that can eject an electron is given by

$$h\nu = E_g. \tag{9.3}$$

The responsivities of several of the most commonly employed photocathode materials are shown in Fig. 9.4. The long wavelength behavior is characteristic of the particular photocathode, while the short wavelength behavior is usually limited by the transmission characteristics of the photomultiplier window. Loci of constant quantum efficiency are shown on these graphs and are related to the responsivity by the relation

$$\mathcal{R}_0 = \frac{\eta e}{h\nu}. \tag{9.4}$$

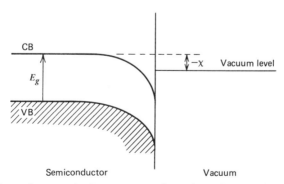

Figure 9.3. Electronic energy-level structure at the surface of a negative-electron-affinity semiconductor.

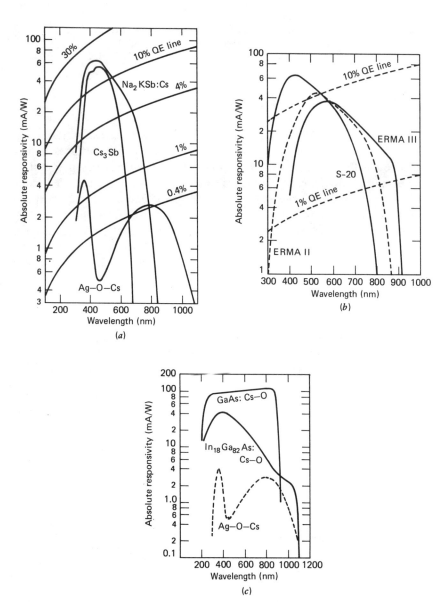

Figure 9.4. Spectral response characteristics of several commercially available photocathodes. (Adapted, with permission, from R. W. Engstrom, *Photomultiplier Handbook*, RCA, Lancaster, Pa., 1980.)

The photocathode labeled Ag—O—Cs in Fig. 9.4a is often designated as S-1 spectral response. The chemical and physical structure of this material is not well understood. An S-1 photocathode is formed by exposing a silver surface to an oxygen glow discharge and to a cesium vapor. Photocathodes comprised of the material Cs_3Sb give rise to spectral responses designated S-4, S-5, S-9, S-11, S-17, S-19, and S-21, depending on the details of the photocathode construction and on the window material. Photomultipliers utilizing Cs_3Sb photocathodes are relatively inexpensive because this material has proven to be robust and easy to manufacture. The final curve is labeled $Na_2KSb:Cs$ and is also known as a tri-alkali or multi-alkali photocathode; its spectral response is often designated S-20.

Two additional spectral response curves labeled ERMA II and III (for extended red multi-alkali) utilizing the $Na_2KSb:Cs$ photocathode are shown in Fig. 9.4b, along with the standard S-20 response. The enhanced red response is achieved primarily by utilizing a thicker layer of photocathode material, and it has the effect of lowering the responsivity at shorter wavelengths. Figure 9.4c shows the S-1 response compared to negative electron affinity photocathodes constructed of GaAs:Cs, O and (InGa)As:Cs, O.

9.2 VACUUM PHOTODIODE

The simplest photoemissive detector is the vacuum photodiode, mentioned briefly in Section 7.1. It can be thought of as a photomultiplier without any amplification stages. Vacuum photodiodes are used to detect strong signals where the greater sensitivity of a photomultiplier is not needed. Since the analysis of the photodiode is simpler than that of the photomultiplier, we shall treat it here and postpone that of the photomultiplier until Section 9.3.

Figure 9.5 shows a vacuum photodiode in which the photocathode and anode are plane parallel plates separated by the distance d. If at time $t = 0$ an electron is ejected from the cathode, it will be accelerated with a force

$$m\ddot{x} = \frac{eV}{d},\tag{9.5}$$

and, neglecting its initial velocity (which is valid so long as its initial kinetic energy $h\nu - \phi$ is much less than the accelerating potential eV), its subsequent velocity will be given by

$$v(t) = \frac{eVt}{md},\tag{9.6}$$

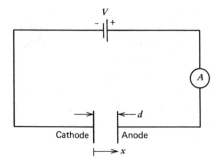

Figure 9.5. Vacuum photodiode with plane-parallel plates.

and its subsequent position will be given by

$$x(t) = \frac{eVt^2}{2md}.$$
(9.7)

The electron will thus reach the anode in a time given by

$$T = d\sqrt{\frac{2m}{eV}}.$$
(9.8)

Since image charges appear on the cathode and anode while the electron is in flight, a current pulse of length T will flow into the external circuit. This current pulse shape is given by

$$i(t) = \frac{ev(t)}{d},$$
(9.9)

or, using the results of Eqs. (9.6)–(9.8), as

$$i(t) = \begin{cases} \dfrac{2et}{T^2} & \text{for } 0 \leqslant t \leqslant T \\ 0 & \text{otherwise,} \end{cases}$$
(9.10)

which is illustrated in Fig. 9.6a.

The frequency response of the photodiode is given by the Fourier transform of $i(t)$, or as

$$I(f) = \int_{-\infty}^{\infty} i(t)e^{-i2\pi ft}\,dt$$

$$= \frac{2e}{(\omega T)^2}\left[(1 + i\omega T)e^{i\omega T} - 1\right],$$
(9.11)

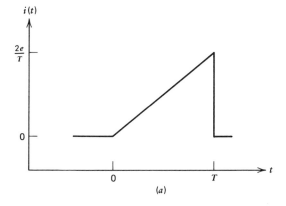

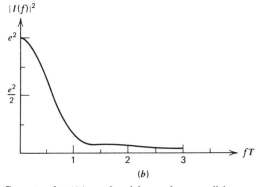

Figure 9.6. (*a*) Current pulse $i(t)$ produced by a plane parallel vacuum photodiode. (*b*) Squared modulus of the Fourier transform of $i(t)$.

where $\omega = 2\pi f$. The squared modulus of this expression is given by

$$|I(f)|^2 = \frac{4e^2}{(\omega T)^4}\left[4\sin^2\left(\frac{\omega T}{2}\right) + (\omega T)^2 - 2\omega T \sin \omega T\right], \quad (9.12)$$

which is shown in Fig. 9.6*b*. The frequency response of the photodiode can thus be expressed as

$$|\mathcal{R}(f)| = \frac{\eta}{h\nu}|I(f)|. \quad (9.13)$$

It is common to characterize the frequency response in terms of a cutoff frequency, defined previously in Eq. (7.8) as the frequency f_c where $|\mathcal{R}(f)|^2$ falls to one-half its maximum value. By inspection of Fig. 9.6b, the cutoff frequency is approximated by

$$f_c \approx \frac{1}{2T}. \qquad (9.14)$$

As an example, if the anode-cathode separation d is taken as 5 mm and the accelerating potential V is taken as 250 V, the pulse duration T is given by Eq. (9.8) as 1 ns, and the cutoff frequency f_c becomes 500 MHz.

The shot noise in the photocurrent can be determined by using Eq. (9.12) for $|I(f)|^2$ in the general result (8.28) for the spectral density. The predominant noise in most vacuum-photodiode applications is, however, the noise in the electronics used to amplify the usually weak signal from the detector. A common method, for instance, of determining the photocurrent leaving the photodiode is to measure the voltage that it develops across a load resistor of large value. Since photodiodes are nearly ideal current sources, large resistances can be used in this manner without decreasing the magnitude of the photocurrent. Under most applications, however, the Johnson noise in the load resistor will limit the NEP of the detector long before the photon-noise limit is reached. The problems of amplifier noise are largely overcome through use of the photomultiplier, discussed in the following section.

9.3 PHOTOMULTIPLIERS

A photomultiplier consists of a photodetector and a low-noise current amplifier combined in a single electron tube. Amplification is achieved by the process of *secondary emission*, in which a single high-energy electron, known as the primary electron, strikes a material surface, causing one or more lower-energy electrons, known as secondary electrons, to be ejected from the surface. The secondary emission ratio δ is defined as the ratio of the average number of secondary electrons leaving the surface to the number of primary electrons incident on the surface. This ratio depends on the kinetic energy of the primary electron, as shown in Fig. 9.7 for several common materials used in photomultipliers. A large value of δ can be obtained with GaP:Cs, which is a negative-electron-affinity material. The other commonly used materials are Cs_3Sb (which is also used to form certain photocathodes) and Cu—BeO—Cs, which is formed by oxidation of a Cu-Be alloy (of ~ 2% Be) and subsequent activation with a surface layer of Cs.

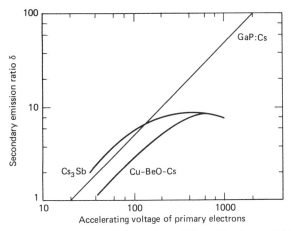

Figure 9.7. Secondary emission ratios for several materials used in photomultipliers. (Adapted, with permission, from R. W. Engstrom, *Photomultiplier Handbook*, RCA, Lancaster, Pa., 1980.)

The operation of a photomultiplier is shown schematically in Fig. 9.8. The secondary-electron emitting surfaces are usually referred to as dynodes and are so labeled in the figure. A high-voltage power supply maintains the photocathode at a negative high voltage, usually in the range 1 to 3 kV, and a voltage divider chain is used to ensure that successive dynodes lie at successively higher potentials. A single photoelectron leaving the photo-cathode is thus accelerated to the first dynode where it causes the ejection of perhaps four secondary electrons each of which is accelerated toward the second dynode where additional electrons are ejected by the same process. Additional amplification takes place at subsequent stages. The gain G of a photomultiplier is defined as the average number of electrons leaving the anode for each photoelectron leaving the photocathode. If each of N dynodes in a photomultiplier is characterized by the same value of δ, the gain of the tube is given by

$$G = \delta^N. \tag{9.15}$$

Typical values of these parameters might be $\delta = 4$, and $N = 9$ implying $G = 3 \times 10^5$. The pulse generated by the initial photoelectron thus gives rise to an anode current i_A which, if desired, can be converted to an output voltage by passing it through a load resistor of value R_L. Alternatively, the anode current can be measured by a true current-measuring device, as shown in Fig. 9.9.

Since the secondary emission ratio δ varies roughly linearly with the accelerating voltage and the gain of an N-stage tube varies as δ^N, it is very

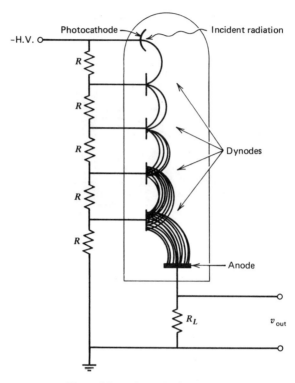

Figure 9.8. Schematic photomultiplier.

important that the operating voltage of a photomultiplier be held fixed. Thus extremely stable power supplies must be used with photomultipliers. If it is important that the anode current depend linearly on the optical power incident on the photocathode, it is also necessary that the current flowing through the tube not be so large as to alter significantly the voltage distribution among the various stages of the tube. A common rule is that the current flowing through the voltage divider chain should be at least 10 times greater than the largest anticipated value of the anode current. Most photomultipliers are designed to produce a maximum of 1 mA of anode current on a continuous basis. When used to detect pulsed radiation, however, a photomultiplier can often produce peak currents a thousand times or more greater than this value. Since it is impractical to pass many tens of amperes through the voltage divider chain to ensure detector linearity under such circumstances, it is common to place capacitors in parallel with the last few resistors of the voltage-divider chain. These capacitors maintain the voltage between stages at a constant value for the

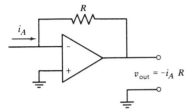

Figure 9.9. Circuit for converting the anode current i_A into an output voltage through use of an operational amplifier.

$v_{out} = -i_A R$

duration of the pulse. Another limitation to the linearity of photomultipliers is the possibility of space-charge buildup inside the tube. This space charge limits the amount of current that can pass through the last few amplification stages.

Noise in Photomultipliers

The two dominant noise sources in a photomultiplier are shot noise in its dark current and noise introduced by the amplification process.

Dark current is the current produced by a photomultiplier when no light is incident on the photocathode. Since dark current usually has the same spectral density as the signal current, there is no electrical way to distinguish the two; thus fluctuations in the dark current constitute noise in the detection process.

At room temperature, the largest source of dark current is usually thermionic emission from the photocathode. If the photocathode is cooled to decrease the amount of thermionic emission, other sources of dark current become important. These include electron emission due to cosmic ray events, the natural radioactivity of the glass envelope of the tube, and the possible ionization of any residual gas within the tube.

For a metallic photocathode of work function ϕ, the rate at which electrons are thermally ejected from the surface is given (for a derivation, see C. Kittel, *Introduction to Solid State Physics*, 1966, p. 246) by the Richardson–Dushman equation

$$i_D = 4\pi m e (kT)^2 h^{-3} A e^{-\phi/kT}, \qquad (9.16)$$

where i_D is the dark current leaving a photocathode of area A. If i_D is measured in amperes, T in degrees Kelvin and A in square centimeters, this equation becomes

$$i_D = 120 A T^2 e^{-\phi/kT}. \qquad (9.17)$$

Probably due to the effect of surface impurities, this equation is found to be only approximately correct for most metal surfaces. It can also provide an order-of-magnitude estimate of the dark current from a semiconductor photocathode. For example, a photocathode of area 1 cm^2 and whose long-wavelength cutoff is 1 μm, implying a work function of $\phi = 1.24$ eV, will produce at $T = 300$ K a dark current of $\sim 2 \times 10^{-14}$ A.

The limitation imposed on the NEP by dark current can be calculated by noting that a current of magnitude i_D appears to have been produced by a radiation background whose power is given by

$$(P_B)_{\text{eff}} = \frac{i_D h\nu}{\eta e}.$$

The standard result (8.60) for the background-noise-limited NEP thus gives

$$P_N = \sqrt{\frac{2h\nu(P_B)_{\text{eff}}\Delta f}{\eta}}$$

$$= \frac{h\nu}{\eta}\sqrt{\frac{2i_D\Delta f}{e}}. \tag{9.18}$$

Assuming that $\Delta f = 1$ Hz, $\eta = 0.1$, $h\nu = 1.24$ eV, and $i_D = 2 \times 10^{-14}$ A (calculated earlier), the NEP becomes 10^{-14} W.

Our discussion has implicitly assumed that the electron multiplication process is noiseless, so that the current signal-to-noise ratio measured at the anode is equal to this ratio calculated at the photocathode. In fact, the electron amplification process is slightly noisy. This noise results from the fluctuations in the size of the current pulse produced by a single photoevent. These fluctuations result from the fact that the actual number of secondary electrons created in an electron impact is a stochastic quantity. Under the assumption that the variance of the number of secondary electrons at any stage is equal to the mean number of secondary electrons δ (as would be the case if the probability distribution were the Poisson distribution), and under the assumption that $\delta \gg 1$ for each stage, the electron amplification process has been shown (see, e.g., Appendix G of the book by Engstrom listed in the bibliography) to decrease the current signal-to-noise ratio by the quantity

$$\gamma \equiv \left(\frac{\delta_1}{\delta_1 - 1}\right)^{1/2}, \tag{9.19}$$

where δ_1 is the secondary emission ratio for the first amplification stage. Thus, under the stated conditions, only the noise introduced by the first dynode is appreciable. The decrease in the signal-to-noise ratio is quite small under these conditions: for $\delta_1 = 5$, the factor is 1.12; for $\delta_1 = 10$, it is 1.05.

A general expression for the noise in the photomultiplier output current (i.e., anode current) can be obtained from the previous results. If radiation of frequency ν and power P falls onto the photocathode, the total cathode current, including dark current i_D is given by

$$i_C = i_D + \frac{\eta e P}{h\nu}, \tag{9.20}$$

and the noise in this current is given, for low frequencies, by Schottky's formula (8.42) as

$$\overline{i_{N,C}^2} = 2ei_C\Delta f. \tag{9.21}$$

Denoting the gain of the electron multiplication process as G, the anode current is then given by

$$i_A = Gi_C. \tag{9.22}$$

Finally, the mean-square noise in the anode current is increased relative to that in the cathode current by a factor of G^2 due to the gain of the tube and by a factor of γ^2 due to the noise in this amplification process. Thus the mean-square noise can be expressed as

$$\overline{i_{N,A}^2} = 2\gamma^2 G^2 ei_C\Delta f, \tag{9.23}$$

or, in terms of the anode current, as

$$\overline{i_{N,A}^2} = 2\gamma^2(Ge)i_A\Delta f. \tag{9.24}$$

Figure 9.10 shows the configuration of the RCA 1P21 photomultiplier. This design and the closely related variations such as the 931A and 1P28 are commonly used in many laboratory detection situations, and some of their operating characteristics (taken from the RCA Photomultiplier Catalog) are outlined here. The 1P21 utilizes a Cs_3Sb photocathode and its spectral response extends from 320 to 600 nm. Its responsivity is maximum at 400 nm and is typically 42 mA/W, corresponding to a quantum efficiency of 12%. Amplification is accomplished by 9 Cs_3Sb dynodes used in a "circular

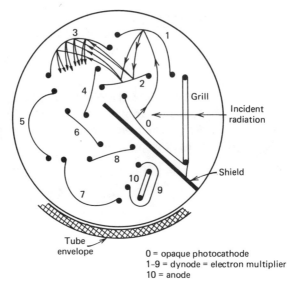

0 = opaque photocathode
1-9 = dynode = electron multiplier
10 = anode

Figure 9.10. The circular-cage photomultiplier structure. (Adapted, with permission, from R. W. Engstrom, *Photomultiplier Handbook*, RCA, Lancaster, Pa., 1980.)

cage" configuration illustrated in the figure. The tube is typically operated at a voltage of 1000 V, leading to a typical gain of 3×10^6, implying an anode responsivity of 1×10^5 A/W. The anode dark current under these conditions is typically 6 nA. It can produce output pulses with rise times as short as 2 ns.

BIBLIOGRAPHY

R. W. Engstrom, *Photomultiplier Handbook*, RCA, Lancaster, Pa., 1980.

H. R. Zwicker, in *Optical and Infrared Detectors*, R. J. Keyes, ed., Springer, Berlin, 1977.

PROBLEMS

1   Assuming that the Richardson–Dushman equation properly predicts the cathode dark current of a photomultiplier whose cutoff wavelength is 1 μm, determine the temperature at which the tube must be operated in order for the dark current to be reduced by a factor of 10 from its room-temperature value.

2   Derive an expression for the additional noise introduced into the output
    of a photomultiplier because of the statistics of the amplification pro-
    cess. Assume that the tube has n stages, that the average gain of the ith
    stage is δ_i, and that the variance of the gain of the ith stage is equal to δ_i
    (as it would be for a Poisson process). Simplify your result when $\delta_i \gg 1$
    for all i.

3   Consider a vacuum photodiode whose sensitivity is limited by Johnson
    noise in its load resistor. Calculate the NEP for a 1-Hz electrical
    bandwidth for Johnson-noise-limited detection as a function of R.
    Evaluate this expression numerically for $R = 10^9 \ \Omega$. What value of the
    photocurrent does this correspond to? What is the rms noise voltage for
    this case?

    At what signal power level would shot noise in the photocurrent just
    equal the Johnson current noise in the load resistor? Evaluate this
    numerically for $R = 10^9 \ \Omega$.

4   A photomultiplier of gain G is used with a room-temperature load
    resistor of value R. If the combined rate at which photoevents and dark
    events occur is denoted r, under what conditions will Johnson noise in
    the load resistor make an appreciable contribution to the noise voltage
    of the system? Evaluate your result numerically for $G = 10^6$ and $R = 1$
    MΩ.

10

Photoconductive Detectors

This chapter discusses (1) the theory of photoconductivity in semiconductors, (2) how this process can be utilized as a detector of optical radiation, and (3) the noise mechanisms that limit the sensitivity of such a device.

10.1 PHOTOCONDUCTIVITY

It has been found that the electrical conductivity of many semiconductors increases greatly when the semiconductor is exposed to optical radiation. This process can be visualized using Fig. 10.1 for the case of an intrinsic (undoped) semiconductor. A photon of energy $h\nu$ greater than the band-gap energy E_g can be absorbed by the semiconductor, thus exciting an electron from the valence band to the conduction band. In many ways, the unoccupied electron state left behind in the valence band can be considered a material particle of charge $+e$ (where the electron charge is $-e$), and this unoccupied electron state is known as a hole. The electron and hole can drift under the influence of an external electric field, giving rise to an electric current. The electron and hole drift velocities are given by the equations

$$\mathbf{v}_n = -\mu_n \mathbf{E}, \tag{10.1a}$$

$$\mathbf{v}_p = \mu_p \mathbf{E}, \tag{10.1b}$$

respectively. Here $\mathbf{E}$ is the applied electric field strength and the quantities μ_n and μ_p are the mobilities for the negative and positive carriers, respectively. In semiconductors at room temperature, the mobilities are typically in the range 10^2 to 10^4 cm^2/V sec, with μ_p often being a factor of 10 smaller than μ_n for a given material. The current density due to this motion of the charges is given by

$$\mathbf{j} = n_p e \mathbf{v}_p - n_n e \mathbf{v}_n, \tag{10.2}$$

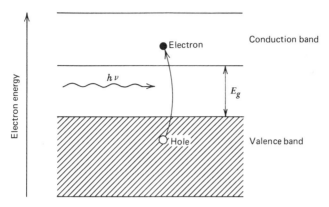

Figure 10.1. Optical absorption in an intrinsic photoconductor. Electrons are shown as closed circles, holes as open circles, and filled bands are shaded.

where n_p and n_n denote the number densities of holes and free electrons, respectively. The current density can alternatively be expressed as

$$\mathbf{j} = \sigma\mathbf{E}, \qquad (10.3)$$

where the electrical conductivity is given by

$$\sigma = n_p e\mu_p + n_n e\mu_n. \qquad (10.4)$$

It should be noted that since electrons and holes drift in opposite directions in the applied field $\mathbf{E}$, they both give rise to a current in the same direction.

The band-gap energies of a number of common semiconductor materials are given in Table 10.1. For use as an intrinsic photoconductive detector, it is necessary to choose a semiconductor whose energy gap E_g is less than the photon energy $h\nu_s$ of the radiation field to be detected. In addition, it is desirable that the band-gap energy E_g nearly equal $h\nu_s$, thus minimizing the background signal resulting from radiation of frequency less than ν_s. One method for matching the semiconductor characteristics to those of the radiation to be detected is to employ solid solutions of two different semiconductors. The band-gap energies of such mixed crystals often vary nearly linearly with the concentration of either component. Two common examples of such systems are $Pb_{1-x}Sn_xTe$ and $Hg_{1-x}Cd_xTe$. It has been established empirically by Schmit and Selzer [J. L. Schmit and E. L. Selzer, *J. Appl. Phys.*, **40**, 4865, (1969)] that the band-gap energy (in eV) of $Hg_{1-x}Cd_xTe$ varies with temperature T (in degrees Kelvin) and the con-

Table 10.1. **Band Gap Energies and Cutoff Wavelengths for
Several Pure Semiconductors at Room Temperature.**

Semiconductor	E_g (eV)	λ_c (μm)
AgCl	3.2	0.39
CdS	2.42	0.51
CdSe	1.74	0.85
CdTe	1.45	0.71
GaAs	1.4	0.88
GaP	2.25	0.55
Ge	0.67	1.8
InAs	0.33	3.7
InSb	0.23	5.4
Si	1.14	1.1
PbS	0.35	3.5
PbSe	0.27	4.6
PbTe	0.30	4.1

centration parameter x as

$$E_g = -0.25 + 1.59x + 0.327x^3 + 5.233 \times 10^{-4}T(1 - 2.08x). \quad (10.5)$$

This relationship is plotted in Fig. 10.2 for $T = 77$ K. For $x < 0.14$, the band-gap energy is negative, thus indicating that the alloy is a semimetal and hence contains free carriers even in the absence of thermal excitation and photoexcitation. $Hg_{1-x}Cd_xTe$ is commonly used as an infrared detector material for x in the range 0.18 to 0.40, corresponding to cutoff wavelengths in the range 3 to 30 μm. In principle, it should be possible to fabricate detectors with an even longer cutoff wavelength by allowing x to approach the value ≈ 0.14, where E_g becomes zero. It has proven difficult, however, to produce materials in which the concentration is sufficiently uniform within the crystal to allow the practical use of such values of x.

Extrinsic Photoconductors

Another method of controlling the spectral response of a photoconductor is through impurity doping. If the impurity has more valence electrons than are needed to bond it to the crystal lattice, the impurity is called a donor impurity because it can become ionized and thereby donate one or more electrons to the conduction band, leading to an increased electrical conductivity. Conversely, an impurity having too few valence electrons to bond it

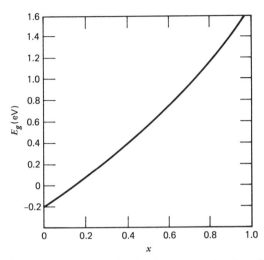

Figure 10.2. Band-gap energy of $Hg_{1-x}Cd_xTe$ at 77 K as a function of the concentration parameter x, as reported by J. L. Schmit and E. L. Seltzer, *J. Appl. Phys.*, **40**, 4865 (1969).

to the crystal lattice is called an acceptor impurity, because it can accept an electron from the valence band to complete its chemical bonds. The hole left behind in the valence band leads to an increased electrical conductivity. It is often convenient to think of this process entirely in terms of the motion of holes. In this picture, an acceptor impurity in its neutral state has a hole bound to it. The process of ionization then corresponds to the hole leaving the impurity and moving down into the valence band.

As shown schematically in Fig. 10.3, donor levels typically lie within the forbidden band close to the bottom of the conduction band, indicating that the donor electron is weakly bound. Similarly, acceptor levels typically lie within the forbidden band close to the top of the valence band, indicating that the acceptor hole is weakly bound. Typical impurity ionization energies lie in the range 0.01 to 0.5 eV. Thus at room temperature and above, such impurities are largely ionized, often leading to significant increases in the electrical conductivity of the semiconductor. At sufficiently low temperatures, however, the impurities are essentially un-ionized, and under such conditions the material will show extrinsic photoconductivity. A photon of energy $h\nu$ greater than the donor ionization energy can, for instance, ionize a donor impurity, thus exciting an electron to the conduction band. Examples of extrinsic photoconductors include germanium doped with gold, mercury, cadmium, copper, zinc, or gallium whose cutoff wavelengths are approximately 5.0, 10, 19, 21, 37, and 120 μm, respectively.

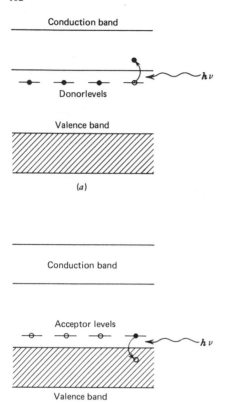

Figure 10.3. Illustrates n-type (a) and p-type (b) extrinsic photoconductivity. Electrons are shown as closed circles, holes as open circles, and filled bands are shaded.

10.2   MODEL PHOTOCONDUCTIVE DETECTOR

Figure 10.4 shows a schematic photoconductive detector. For simplicity, it will be assumed that the detector is either an n-type extrinsic photoconductor or an intrinsic photoconductor with $\mu_n \gg \mu_p$, so that in either case the motion of only the electrons need be considered. We suppose that monochromatic radiation of frequency ν and power P falls onto the detector and that a fraction η of this power, when absorbed, leads to the excitation of electrons to the conduction band.

The electrical properties of the photoconductor can be described in terms of the surface electron density S of conduction-band electrons; S is thus measured in units of electrons per unit area of the detector surface. We shall denote by ΔS the contribution to S due to photo-excited electrons; thermally excited electrons may also contribute to S. If it assumed that τ is the

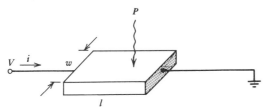

Figure 10.4. Schematic photoconductive detector.

mean lifetime of an electron in the conduction band, the time evolution of ΔS is governed by

$$\dot{\Delta S} = \frac{\eta P}{h\nu wl} - \frac{\Delta S}{\tau}, \qquad (10.6)$$

which has the steady-state solution

$$\Delta S = \frac{\eta P\tau}{h\nu wl}. \qquad (10.7)$$

If a bias voltage V is applied across the photoconductor, each electron will by Eq. (10.1a) acquire a drift velocity

$$v_n = -\mu_n E, \qquad (10.8)$$

where the electric field strength is given as $E = V/l$. The photocurrent thus induced to flow is given by

$$i = -\Delta Sev_n w$$

$$= \frac{\eta ePG}{h\nu}, \qquad (10.9)$$

where G is known as the photoconductive gain and can be expressed in either of the forms

$$G = \frac{|v_n|\tau}{l}$$

$$= \frac{\mu\tau v}{l^2}. \qquad (10.10)$$

The photoconductive gain can be interpreted as the average number of electrons that flow through the circuit for each photoevent. Since G is linearly proportional to τ, the desirable detector properties of having a large gain and a fast response are generally incompatible. Photoconductive gains as large as 1000 have been measured. Values of G greater than unity require the flow of more than one electron through the detector for each photoevent. This is possible because conduction-band electrons can leave the photoconductor at the positive electrode and be replaced by additional electrons entering the photoconductor from the negative electrode. This process is terminated only when the electron recombines with an excited donor. [It is assumed here that the metal contacts to the photoconductor are ohmic (i.e., not rectifying), thus allowing a free flow of current between the metal and photoconductor.] In contrast, the vacuum photodiode discussed in Chapter 9 allows exactly one electron to flow through the circuit for each photoevent, since electrons can leave the negative electrode only by photoemission.

10.3   NOISE MECHANISMS IN PHOTOCONDUCTORS

Figure 10.5 shows schematically the current-noise power spectrum of a photoconductor carrying a current. Three distinct noise mechanisms that commonly occur in semiconductors are shown; each can dominate in certain frequency ranges.

At low electrical frequencies, a mechanism often called $1/f$ noise is usually dominant. The origin of this noise source is not well understood. It has, however, been empirically established that its contribution to the current noise is of the form

$$\overline{i_N^2}(f) = \frac{Ki^\alpha}{f^\beta}, \tag{10.11}$$

where K is a proportionality factor and where $\alpha \simeq 2$ and $\beta \simeq 1$. The effects of $1/f$ noise can be minimized by chopping the signal radiation at some

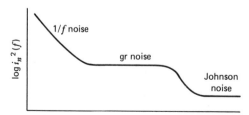

Figure 10.5.   Typical semiconductor current-noise power spectrum.

large frequence (10^3 Hz is often sufficiently fast) and examining only the frequency components of the photocurrent contained in a narrow frequency band including the chopping frequency. This operation can be conveniently performed through the use of a lock-in amplifier.

The power spectrum of Johnson noise is flat at all commonly encountered electrical frequencies; Johnson noise thus dominates at frequencies higher than the cutoff frequencies for other noise mechanisms. Johnson noise arises from the thermal agitation of electrons and holes in the semiconductor. Its current-noise power spectrum is given by Eq. (8.66) as

$$\overline{i_N^2}(f) = \frac{4kT}{R},$$

R being the resistance of the photoconductor.

A noise mechanism peculiar to semiconductors is generation-recombination (gr) noise. Generation-recombination noise results from the statistical fluctuations in the number of conduction-band electrons and valence-band holes that are available to conduct electricity at any given instant. This number fluctuates because both the generation process and the recombination process are stochastic and in many cases can be considered independent of one another. A consequence of having two independent noise sources is that a photoconductive detector is at least $\sqrt{2}$ times noisier than an ideal photon detector in a sense that will be defined in Section 10.4. A general theory of gr noise in which both thermally and optically excited carriers are present is rather complicated and will be postponed until Section 10.5.

For the present, we shall consider gr noise in a detector at a sufficiently low temperature that only photoexcited free carriers are present. We shall assume, as in the preceding section, that the mean lifetime τ of an electron in the conduction band is a constant. Moreover, we shall assume for simplicity that only the motion of electrons contributes to the current flow (as in the case of an extrinsic, n-type photoconductor or an intrinsic photoconductor with $\mu_n \gg \mu_p$).

The photocurrent passing through the photoconductor will be comprised of pulses whose lengths fluctuate randomly, as shown schematically in Fig. 10.6. Since the duration of a given current pulse is equal to the lifetime of the conduction-band electron, the probability that the duration of the pulse lies within the range t to $t + dt$ is given by

$$p(t)\,dt = \frac{1}{\tau}e^{-t/\tau}\,dt. \qquad (10.12)$$

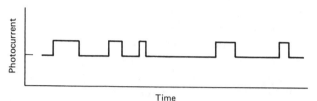

Figure 10.6. Origin of generation recombination noise. The photoinduced current pulses occur at random instants of time and possess random lengths. This effect leads to a $\sqrt{2}$ degradation in the NEP of a background-limited photoconductor relative to that of a detector limited only by shot noise.

The analysis of photoinduced gr noise involves generalizing the analysis of shot noise given previously. The usual shot noise formula (8.42) is given by

$$\overline{i_N^2} = 2e\bar{i}\,\Delta f.$$

This formula can be applied to the calculation of gr noise by noting that the contribution to $\overline{i_N^2}$ from pulses of length t is given by

$$d\left(\overline{i_N^2}\right)2q(t)\,\Delta f\,d\bar{i}, \tag{10.13}$$

where $q(t)$ is the charge passing through the photoconductor for pulses of length t and where $d\bar{i}$ is the contribution to $\bar{i}$ resulting from these pulses. Since the charge per pulse is proportional to the length t of the pulse, and since the average value of this charge is Ge, the quality $q(t)$ must be given by

$$q(t) = \frac{Get}{\tau}. \tag{10.14a}$$

Similarly, the contribution to the mean current resulting from pulses of length t must be proportional to the pulse length t and to the probability distribution $p(t)$ for the pulse length, giving

$$d\bar{i} = \bar{i}\,\frac{t}{\tau^2}e^{-t/\tau}\,dt. \tag{10.14b}$$

The mean-square current noise can now be obtained by integrating Eq. (10.13) over all possible pulse lengths to give

$$\overline{i_N^2} = 2Ge\bar{i}\,\Delta f\int_0^\infty \frac{t^2}{\tau^3}e^{-t/\tau}\,dt$$

$$= 4Ge\bar{i}\,\Delta f. \tag{10.15}$$

This inequality is most readily met if the equivalent resistance R is large, which by Eq. (10.20) requires that both the detector resistance R_D and the load resistance R_L be large.

Let us now assume that the two conditions stated in the last paragraph are met, so that the detector is limited primarily by gr noise in the photocurrent. We can calculate the noise equivalent power (NEP) by calculating the value of P_S which gives a value of $\overline{i_S^2}$ equal to $\overline{i_N^2}$, giving the result

$$P_N = \sqrt{\frac{4P_B h\nu \Delta f}{\eta}}. \tag{10.22}$$

[By comparison with Eq. (8.60) for the NEP of a detector limited by photon noise only, it can be seen that the NEP of a photoconductive detector is increased by a factor of $\sqrt{2}$, illustrating the increased noise when the gr process is present.]

10.5 THEORY OF GENERATION-RECOMBINATION NOISE

This section presents a general discussion of generation-recombination noise (gr noise) in photoconductive detectors. The model presented here allows both thermally generated and photo-generated carriers to be present, and it can be applied to either an intrinsic or extrinsic photoconductor.

We consider a slab of semiconductor of cross sectional area A and length L to which a potential of magnitude V has been applied. The instantaneous value of the current passing through the semiconductor is then related to the total number of free electrons $N(t) = ALn_n(t)$ and the total number of free holes $P(t) = ALn_p(t)$ by

$$i(t) = \frac{eV}{L^2}\left(\mu_n N(t) + \mu_p P(t)\right). \tag{10.23}$$

The mean value of the current is thus related to the mean value of the number of carriers by

$$\bar{i} = \frac{eV}{L^2}\left(\mu_n \overline{N} + \mu_p \overline{P}\right). \tag{10.24}$$

The fluctuation $\Delta i = i(t) - \bar{i}$ in the current flow is thus related to the fluctuations in the carrier concentrations $\Delta N = N(t) - \overline{N}$ and $\Delta P = P(t)$

$- \bar{P}$ by the expression

$$\overline{(\Delta i)^2} = \frac{e^2 V^2}{L^4} \left(\mu_n^2 \overline{(\Delta N)^2} + 2\mu_n \mu_p \overline{(\Delta N)(\Delta P)} + \mu_p^2 \overline{(\Delta P)^2} \right), \quad (10.25)$$

which can be expressed in terms of $\bar{i}$ and the ratio

$$b \equiv \frac{\mu_n}{\mu_p} \qquad (10.26)$$

as

$$\overline{i_{gr}^2} \equiv \overline{(\Delta i)^2} = \left(\frac{\bar{i}}{b\bar{N} + \bar{P}} \right)^2 \left[b^2 \overline{(\Delta N)^2} + 2b \overline{(\Delta N)(\Delta P)} + \overline{(\Delta P)^2} \right].$$

$$(10.27)$$

(In the remainder of this section, we shall use the symbol $\overline{i_{gr}^2}$ to denote the mean-square current noise, previously denoted $\overline{i_N^2}$, to avoid any possible confusion of the subscript N with the number of free electrons N). Similarly, the spectral density of the current fluctuations is given by

$$\overline{i_{gr}^2}(f) = \left(\frac{\bar{i}}{b\bar{N} + \bar{P}} \right)^2 \left[b^2 \overline{(\Delta N)^2}(f) + 2b \overline{(\Delta N)(\Delta P)}(f) + \overline{(\Delta P)^2}(f) \right].$$

$$(10.28)$$

From Eqs. (10.27) and (10.28), the current noise can be calculated from a knowledge of the fluctuations in N and P.

In the most general circumstances, it is very difficult to calculate the quantities $\overline{(\Delta N)^2}$, $\overline{(\Delta P)^2}$ and $\overline{(\Delta N)(\Delta P)}$. Considerable simplification results, however if it is possible to make a two-level approximation, so that the statistical properties of both N and P can be described in terms of the properties of a single random variable, say N. One example of such a system is an intrinsic photoconductor containing no impurity levels, so that $\Delta P(t) = \Delta N(t)$ at all times. Another example is an extrinsic photoconductor (say n-type) at a temperature sufficiently low that P, and hence ΔP, are zero at all times, and thus only $N(t)$ may fluctuate. In any case, we shall assume that the statistical properties of the medium can be described solely in terms of the number of free electrons $N(t)$.

The approach that we shall adopt in describing the statistical properties of $N(t)$ is based on a detailed balancing of the generation and recombination rates. This method was originally introduced by Burgess (references to this work are included at the end of this chapter). We define the generation rate $g(N)$ such that, if N electrons are already in the conduction band, $g(N)\,dt$ denotes the probability that an additional electron is excited to the conduction band in the time interval dt. Both thermal excitation and photoexcitation contribute to this generation rate. Similarly, the recombination rate $r(N)$ is defined such that $r(N)\,dt$ denotes the probability that an electron leaves the conduction band in time dt. If it is assumed that the functional forms $g(N)$ and $r(N)$ are known, a *master equation* can be developed for the probability distribution function $p(N)$, which gives the probability that the conduction band contains N electrons:

$$\frac{dp(N)}{dt} = r(N+1)p(N+1) + g(N-1)p(N-1)$$

$$- [g(N) + r(N)]\, p(N). \tag{10.29}$$

In steady state, it must be true that $dp(N)/dt = 0$ and Eq. (10.29) can be solved iteratively to obtain an expression for $p(N)$:

$$\frac{p(N)}{p(0)} = \prod_{\nu=0}^{N-1} g(\nu) \bigg/ \prod_{\nu=1}^{N} r(\nu). \tag{10.30}$$

The most probable value $\bar{N}$ of N can be obtained by treating N as a continuous variable and setting the derivative of $p(N)$ with respect to N equal to zero. This procedure is equivalent to setting the logarithmic derivative of p equal to zero:

$$0 = \frac{d}{dN} \ln p(N) \bigg|_{N=\bar{N}} = \ln g(\bar{N}) - \ln r(\bar{N}+1). \tag{10.31}$$

Since $\bar{N} \gg 1$, this equilibrium condition is equivalently expressed as

$$g(\bar{N}) = r(\bar{N}). \tag{10.32}$$

In order to determine the variance in N, we first show that we can approximate $p(N)$ by a Gaussian centered on $\bar{N}$. We note that

$$\frac{d^2}{dN^2} \ln p(N) \bigg|_{N=\bar{N}} = \frac{g'(\bar{N})}{g(\bar{N})} - \frac{r'(\bar{N})}{r(\bar{N})}, \tag{10.33}$$

where the prime denotes differentiation with respect to N. We can thus express $\ln p(N)$ as a power series expansion centered on $N = \overline{N}$:

$$\ln p(N) = \ln p(\overline{N}) - \tfrac{1}{2}(N - \overline{N})^2 \left[\frac{r'(\overline{N})}{r(\overline{N})} - \frac{g'(\overline{N})}{g(\overline{N})}\right], \quad (10.34)$$

and thus

$$p(N) = p(\overline{N})e^{-(N-\overline{N})^2/2\overline{(\Delta N)^2}}, \quad (10.35)$$

where the variance of N is given by

$$\overline{(\Delta N)^2} = \frac{g(\overline{N})}{r'(\overline{N}) - g'(\overline{N})}. \quad (10.36)$$

The significance of the denominator of this expression can be determined by considering the relaxation of the system to equilibrium after it has been disturbed. The expected time evolution of the system, after it has been disturbed by an amount $N - \overline{N}$, is given by

$$\frac{d}{dt}\overline{(N - \overline{N})} = \frac{\overline{dN}}{dt} = g(N) - r(N)$$

$$\simeq [r'(\overline{N}) - g'(\overline{N})](N - \overline{N}).$$

The system thus relaxes the equilibrium according to

$$\Delta N(t) = \Delta N(0)e^{-t/\tau'}, \quad (10.37)$$

where the relaxation time is given as

$$\tau' = \frac{1}{[r'(\overline{N}) - g'(\overline{N})]}. \quad (10.38)$$

The quantity τ' can also be interpreted as the response time of the photodetector or as the incremental carrier lifetime, that is, the lifetime of an additional carrier, when N electrons are known to be already present. τ' will not in general be equal to the mean lifetime $\bar{\tau}$ of a thermally or optically generated carrier (often called the excess carrier lifetime, since it is the mean lifetime of carriers in excess to those present when $g = 0$), which is given by

$$\bar{\tau} = \frac{\overline{N} - \overline{N}(g = 0)}{g(\overline{N})}. \quad (10.39)$$

The result (10.38) can be used to express Eq. (10.36) for the variance in N as

$$\overline{(\Delta N)^2} = g(\overline{N})\tau'. \qquad (10.40)$$

This result was first obtained by Burgess and is known as the generation-recombination theorem. Using Eqs. (10.38)–(10.40) and the previous result (8.7), we can conclude that the correlation function of N is given by

$$C_N(t) = g(\overline{N})\tau' e^{-|t|/\tau'}.$$

The spectral density of N is thus obtained using the Wiener–Khintchine relation (8.17) as

$$\overline{(\Delta N)^2}(f) = \frac{4g(\overline{N})\tau'^2}{1 + (2\pi f \tau')^2}$$

or as

$$\overline{(\Delta N)^2}(f) = 4g(\overline{N})\tau'^2 \quad \text{for } f \ll 1/\tau'. \qquad (10.41)$$

Intrinsic Photoconductor

The general results just deduced will now be applied to several special cases. The first example is an intrinsic photoconductor, shown in Fig. 10.8, for which case the number of free holes is equal to $P(t) = N(t) - N_d + N_a$, where N_d denotes the number of donor atoms and N_a denotes the number of acceptor atoms, both of which are assumed to be totally ionized. It is necessary to assume that the impurities are *totally* ionized so that the number of ionized impurities does not fluctuate and thus that the two-level model just discussed is applicable.

Since in any reasonable situation, N is much less than the total number of bound electrons in the valence band, it can be assumed that $g(N)$ is independent of N for either thermal excitation or photoexcitation. Thus

$$g(N) = g_0. \qquad (10.42)$$

The generation rate g_0 will have both a thermal contribution and a contribution that depends linearly on the incident photon flux. If the recombination process involves a direct band-to-band transition, the recombination coefficient is given by

$$r(N) = \rho NP = \rho N(N - N_d - N_a) \qquad (10.43)$$

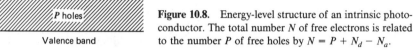

Conduction band

N electrons

——— N_d ionized donor

——— N_a ionized acceptors

P holes

Valence band

Figure 10.8. Energy-level structure of an intrinsic photoconductor. The total number N of free electrons is related to the number P of free holes by $N = P + N_d - N_a$.

for some constant ρ. The relaxation time is then given by Eq. (10.38) as

$$\tau' = \frac{1}{\rho(\overline{N} + \overline{P})}, \tag{10.44}$$

where the equilibrium carrier concentrations $\overline{N}$ and $\overline{P}$ are related according to Eq. (10.32) by

$$g_0 = \rho\overline{N}\overline{P} = \rho\overline{N}(\overline{N} - N_d + N_a). \tag{10.45}$$

The mean-square fluctuation in carrier number is then given by Eq. (10.40) as

$$\overline{(\Delta N)^2} = g_0\tau' = \frac{\overline{N}\overline{P}}{\overline{N} + \overline{P}}, \tag{10.46}$$

and the spectral density of these fluctuations are given by Eq. (10.41) as

$$\overline{(\Delta N)^2}(f) = 4\tau'\frac{\overline{N}\overline{P}}{\overline{N} + \overline{P}} \tag{10.47}$$

for $f \ll 1/\tau'$. Finally, the mean-square current noise in an electrical bandwidth Δf can be calculated from Eq. (10.28), using the relation $\Delta P = \Delta N$, as

$$\overline{i_{gr}^2} = 4\bar{i}^2\left(\frac{b + 1}{b\overline{N} + \overline{P}}\right)^2\frac{\overline{N}\overline{P}}{\overline{N} + \overline{P}}\tau'\Delta f. \tag{10.48}$$

A special case of the intrinsic photoconductor is obtained by assuming in addition that the material is an intrinsic *semiconductor*, that is that $N_d = N_a$ so that $N = P$ at all times. This case can be treated by setting $\overline{N}$ equal to $\overline{P}$ in Eqs. (10.42) through (10.48).

Extrinsic Photoconductors

A second example of the application of the general results of this section is afforded by the extrinsic photoconductor. For definiteness, we assume that it is an *n*-type extrinsic photoconductor, as shown in Fig. 10.9. The photoconductor is assumed to possess N_d donors, of which N are ionized leading to free carriers in the conduction band. The number P of free (i.e., valence band) holes is assumed to be negligibly small. The generation rate is assumed to be proportional to the number of un-ionized donors so that

$$g(N) = \gamma(N_d - N), \qquad (10.49)$$

where the generation coefficient γ includes a thermal contribution and a radiative contribution that is linear in the incident photon flux. The recombination rate is assumed to be proportional to the product of the number of free electrons and to the number of ionized donors so that

$$r(N) = \rho N^2 \qquad (10.50)$$

where ρ is the recombination coefficient. The equilibrium number of conduction band electrons is thus given according to Eq. (10.32) by

$$\overline{N} = -\frac{1}{2}\frac{\gamma}{\rho} + \left[\frac{1}{4}\left(\frac{\gamma}{\rho}\right)^2 + \frac{\gamma}{\rho}N_d\right]^{1/2}. \qquad (10.51)$$

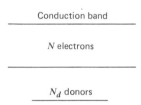

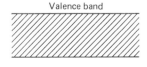

Figure 10.9. Energy-level structure of an *n*-type extrinsic photoconductor. Of the N_d donors, N are ionized and $(N_d - N)$ are un-ionized.

The relaxation time is then given by Eq. (10.38) as

$$\tau' = \frac{1}{\gamma + 2\rho\bar{N}} = \frac{1}{\rho N_d} \frac{1 - \lambda}{\lambda(2 - \lambda)}, \tag{10.52}$$

where

$$\lambda \equiv \frac{\bar{N}}{N_d} \tag{10.53}$$

denotes the fractional ionization of the impurity. The variance of N is given by Eq. (10.40) as

$$\overline{(\Delta N)^2} = \frac{\bar{N}(N_d - \bar{N})}{2N_d - \bar{N}} = \bar{N}\left(\frac{1 - \lambda}{2 - \lambda}\right) = N_d \frac{\lambda(1 - \lambda)}{2 - \lambda}. \tag{10.54}$$

This variation of $\overline{(\Delta N)^2}$ with λ is illustrated in Fig. 10.10. Finally, the mean-square current noise in bandwidth Δf is given in the limit $f \ll 1/\tau'$ by Eq. (10.27) as

$$\overline{i_{gl}^2} = 4\frac{\bar{i}^2}{N}\left(\frac{1 - \lambda}{2 - \lambda}\right)\tau'\Delta f. \tag{10.55}$$

Inspection of Eqs. (10.48) and (10.55) reveals that, for the cases considered here, the gr current noise does not in general follow the simple relation

$$\overline{i_{gr}^2} = 4Ge\bar{i}\,\Delta f,$$

derived in Section 10.2 for an idealized photoconductor characterized by a *constant* carrier lifetime τ and no dark current. The origin of the difference is twofold: (1) For the intrinsic photoconductor, we have allowed the possibility that $\bar{N} \neq \bar{P}$. This could occur, for instance, if the material contains some totally ionized acceptors. The free holes thus created would of course contribute to the electrical conductivity, but they would not produce any gr noise. (2) For both cases treated here, the relaxation time τ' is not a constant but depends on the number of free carriers and thus on the generation rate. As a result, individual contributions to the photocurrent become partially correlated, since the size of the current pulse produced by one photoevent depends on the number of free electrons present at that time and thus on the number of photoevents that occurred in the immediately

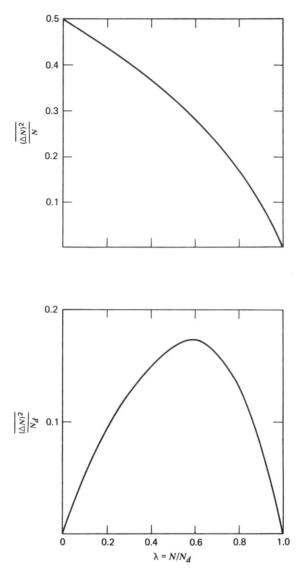

Figure 10.10. Mean-square fluctuation in the number of free carriers in an extrinsic photoconductor, as given by Eq. (10.54).

preceding time interval. As a result, it is not in general true that $\overline{(\Delta N)^2} = \overline{N}$, as is the case for fluctuations described by the Poisson probability distribution. It should be noted that conditions do exist under which the relaxation time τ' is independent of the free carrier concentration, so that the simple analysis of Section 10.3 can become valid (so long as $\overline{N} = \overline{P}$ for an intrinsic photoconductor, as mentioned as point (1) above). In particular, if the generation rate is independent of N, (e.g., $g(N) = g_0$) and if the recombination rate depends linearly on the carrier concentration (e.g., $r(N) = \beta N$), the response time τ' and mean excess carrier lifetime $\bar{\tau}$ will be constant and both equal to β^{-1}. The mean-square fluctuation in N will then be given by Eq. (10.40) as $\overline{(\Delta N)^2} = \overline{N}$. For an intrinsic photoconductor, constant lifetimes can occur if the recombination mechanism involves transitions to recombination centers that are much more numerous than the number of free electrons. Recombination by this method has been treated by Schockley and Read [W. Schockley and W. T. Read, Jr., *Phys. Rev.*, **87**, 835 (1952)]. Similarly, constant lifetimes can occur for an extrinsic photoconductor if the number of empty donors greatly exceeds the number of free electrons, as could occur in a material containing both donor and acceptor impurities. This possibility has been treated by Burgess [R. E. Burgess, *Proc. Phys. Soc. London Sect. B*, **69**, 1020 (1956)].

Several additional topics exist that cannot be treated in detail here but which can be quite important in the design actual detectors. In general, both photo-generated electrons and holes will drift under the influence of an applied electric field. For sufficiently large field strengths, both types of carriers can drift to their respective electrodes in a mean carrier lifetime. This effect is known as ambipolar sweepout and has been shown to limit both the photoconductive gain and the gr noise. Ambipolar sweepout has been treated by Rittner (E. S. Rittner, in *Photoconductivity Conference*, R. G. Breckenridge, ed., Wiley, New York, 1956) and its influence on the performance of infrared detectors has been discussed by Williams [R. L. Williams, *Infrared Phys.*, **8**, 337 (1968)], by Johnson (M. R. Johnson, *J. Appl. Phys.*, **43**, 3090 (1972)] and by Emmons and Ashley [S. P. Emmons and K. L. Ashley, *Appl. Phys. Lett.* **20**, 161 (1972)]. In addition, the theory of gr noise can become quite complicated if one allows the possibility that recombination involves cascades between energy levels lying within the forbidden band or that recombination can occur via three-body collisions, that is, by Auger processes. These recombination processes have been discussed by van Vliet and Fassett (K. M. van Vliet and J. R. Fassett, in *Fluctuation Phenomena in Solids*, R. E. Burgess, ed., Academic, New York, 1965) and by Kinch et al. [M. A. Kinch, M. J. Brau and A. Simmons, *J. Appl. Phys.*, **44**, 1649 (1973)].

BIBLIOGRAPHY

General References

R. Dalven, *Introduction to Applied Solid-State Physics*, Plenum, New York, 1980.

C. Kittel, *Introduction to Solid-State Physics*, Wiley, New York, 1976, especially Chapter 8.

D. W. Kruse, L. D. McGlauchlin, and R. B. McQuistan, *Infrared Technology*, Wiley, New York, 1962.

D. Long, in *Optical and Infrared Detectors*, R. J. Keyes, ed., Springer, Berlin, 1977.

E. S. Rittner in *Photoconductivity Conference*, R. G. Breckenridge, ed., Wiley, New York, 1956.

A. Rose, *Concepts in Photoconductivity and Allied Problems*, Interscience, New York, 1963.

R. A. Smith, F. E. Jones, and R. P. Chasmar, *The Detection and Measurement of Infrared Radiation*, Oxford, University, London, 1968.

S. M. Sze, *Physics of Semiconductor Devices*, Wiley, New York, 1969.

References to the Theory of gr Noise

R. E. Burgess, *Physica*, **20**, 1007 (1954); *Proc. Phys. Soc. London Sect. B*, **68**, 661 (1955); **69**, 1020 (1956).

A. van der Ziel, *Noise*, Prentice-Hall, Englewood Cliffs, N.J., 1970, Chapters 2 and 5, and *Noise in Measurement*, Wiley, New York, 1976, Chapters 5 and 11.

K. M. van Vliet, *Proc. IRE*, **46**, 1004 (1958); *Appl. Opt.*, **6** 1145 (1967).

K. M. van Vliet and J. R. Fassett, in *Fluctuation Phenomena in Solids*, R. E. Burgess, ed. Academic, New York, 1965.

PROBLEMS

1 Explain how to determine both the quantum efficiency and the photo-conductive gain of an infrared photoconductive detector. Assume that you have ability to measure the responsivity and the detector noise and that the simple theory of photoconductivity given in Sections 10.2–10.4 is valid.

2 Generalize the treatment of the extrinsic photoconductor given in Section 10.5 by assuming that M of the N_d donors are ionized even at temperature $T = 0$ K. The generation rate, for instance, is then given by $g(N) = \gamma(N_d - M - N)$. Calculate τ', $\bar{\tau}$, $\overline{(\Delta N)^2}$ and $\overline{i_{gr}^2}$ as a function of M and $\bar{N}$.

3   Calculate the NEP of the intrinsic photoconductive detector treated in
    Section 10.5 as a function of $\bar{N}$ and $\bar{P}$, and describe how $\bar{N}$ and $\bar{P}$
    depend on the temperature of the detector and on the background
    photon flux.

4   Calculate the NEP of the extrinsic photoconductive detector treated in
    Section 10.5 as a function of $\bar{N}/N_d$, and describe how this ratio depends
    on the temperature of the detector and on the background photon flux.

11

Photovoltaic Detectors

Photovoltaic detectors are solid-state devices capable of producing a photo-current or photovoltage when operated without an external electrical bias. The most common example is the semiconductor photodiode, which is often referred to simply as the photodiode when there is no chance of confusion with the vacuum photodiode.

This chapter presents a detailed analysis of the abrupt p-n junction photodiode. This idealization makes the analysis mathematically tractable, and leads to an analysis that properly displays the qualitative and many of the quantitative features of the photovoltaic detectors. Also included in this chapter is a brief discussion of other types of photovoltaic detectors, such as the avalanche photodiode, the p-i-n photodiode, and the Schottky barrier photodiode.

11.1 THE p-n JUNCTION PHOTODIODE

A p-n junction is formed by doping adjacent regions of a semiconductor with donor and acceptor impurities in such a way that a sharp transition between p- and n-type regions is formed. The rectifying properties of a p-n junction that make it useful as a diode also make it useful as a detector of radiation. In the following analysis, it will be assumed for simplicity that the transition from the p- to n-type regions is discontinuous. Such a junction is called an abrupt junction.

The electrical properties of a p-n junction can be understood qualitatively from the energy-band diagrams shown in Fig. 11.1. Two separated pieces of p- and n-type materials are shown in Fig. 11.1a. It can be assumed that at room temperature most of the impurity levels are ionized. Thus there are free holes in the valence band of the p-type material and free electrons in the conduction band of the n-type material. Charge neutrality is maintained at all points in each material, however, because of the presence of the ionized impurities. The Fermi levels in the two materials are located near their respective impurity levels, as shown in the figure.

If the two materials are brought into electrical contact, as shown in Fig. 11.1b, free electrons can diffuse from the n-type to the p-type region, and free holes can diffuse from the p-type to the n-type region. Both processes lead to the recombination of electrons with holes, giving rise to a region that is depleted of its mobile charge. This region is called the depletion layer. This region is also known as the space-charge region because it is not

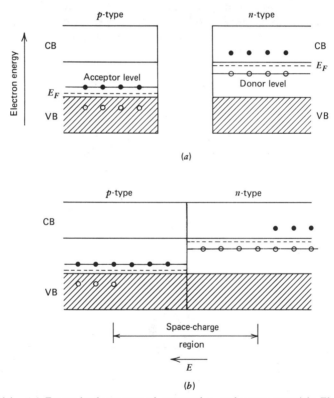

Figure 11.1. (a) Energy-level structure of separated p- and n-type materials. Electrons are represented by closed circles, and holes are represented by open circles. (b) When the materials are brought into electrical contact, electrons diffuse into the p-type region and holes diffuse into the n-type region, creating a space-charge region. (c) The resulting built-in electric field gives rise to a contact potential V_0 between the two regions, raising the energy of the p-type material relative to that of the n-type material. (d) In the presence of an applied reverse voltage, the magnitude of the potential barrier is increased.

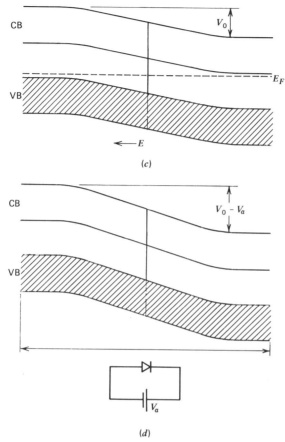

Figure 11.1. (*Continued*)

electrically neutral, since it contains ionized donors and acceptors. These two layers of space charge produce a built-in electric field, which is directed so as to prevent the diffusion of additional free carriers into the depletion layer.

The existence of this built-in electric field creates a potential difference V_0 between the *p*- and *n*-type materials, causing the energies of the bands of the two materials to be shifted, as shown in Fig. 11.1*c*. The condition of equilibrium requires that the Fermi levels of the two materials be equal, which is also illustrated in the figure. This contact potential between the *p*- and *n*-type materials is responsible for the rectifying action of a *p-n* junction. If a voltage is applied between the *p*- and *n*-type materials, most of the voltage drop will occur across the depletion layer, since this region is depleted of mobile charges and thus has a large resistance. The applied voltage is said to be positive if the p-type material is at a higher voltage than the n-type material. A voltage of this sense will tend to cause a current to flow to the right in Fig. 11.1, and this is considered to constitute a positive current. The junction is said to be forward biased when a positive voltage is applied. Conversely, the junction is said to be back biased (or reverse biased) if a negative voltage is applied. Figure 11.1d shows the energy-band structure in the neighborhood of a p-n junction for the case of a negative applied voltage. The height of the potential barrier separating the p- and n-type regions is increased for this sign of the applied voltage.

The nature of the space charge region can be analyzed quantitatively by solving Poisson's equation

$$\nabla^2 V = -\frac{\rho}{\varepsilon}, \tag{11.1}$$

where $\mathbf{E} = -\nabla V$, with appropriate boundary conditions. (The notation used here is similar to that used in the book by Yariv listed at the end of this chapter.) We assume that the junction occupies the plane $x = 0$. We also assume that the number density of ionized acceptors has the constant value N_A for $x < 0$ and that the number density of ionized donors has the constant value N_D for $x > 0$. If we let l_p and l_n be the distances that the depletion layer extends into the p- and n-type regions, respectively, Poisson's equation becomes

$$\frac{d^2V}{dx^2} = \begin{cases} \dfrac{eN_A}{\varepsilon} & -l_p < x < 0, \\[2mm] \dfrac{-eN_D}{\varepsilon} & 0 < x < l_n, \\[2mm] 0 & \text{otherwise.} \end{cases} \tag{11.2}$$

The solutions must obey the condition

$$E = -\frac{dV}{dx} = 0 \quad \text{for } x \leqslant -l_p \text{ and } x \geqslant l_n, \tag{11.3}$$

which ensures that the entire potential drop is across the depletion layer; they must obey the requirement that $V(x)$ and its derivative be continuous at $x = 0$; and they must also obey the condition

$$V(l_n) - V(-l_p) = V_0 - V_a, \tag{11.4}$$

where V_0 is the contact potential between the two regions (which is equal to the energy difference of their Fermi levels) and where V_a is applied voltage. The solution to Eq. (11.2) is then given by

$$
V = \begin{cases}
\dfrac{eN_A}{2\varepsilon}\left(x^2 - 2l_p x\right) & -l_p < x < 0, \\[3mm]
-\dfrac{eN_D}{2\varepsilon}\left(x^2 - 2l_n x\right) & 0 < x < l_n,
\end{cases}
\tag{11.5}
$$

where the distances l_p and l_x are given by

$$
l_p = \left[\left(\frac{2\varepsilon}{e}\right)\left(\frac{N_D/N_A}{N_A + N_D}\right)(V_0 - V_a)\right]^{1/2},
\tag{11.6}
$$

$$
l_n = \left[\left(\frac{2\varepsilon}{e}\right)\left(\frac{N_A/N_D}{N_A + N_D}\right)(V_0 - V_a)\right]^{1/2}.
\tag{11.7}
$$

These distances thus obey the relation

$$
N_A l_p = N_D l_n,
\tag{11.8}
$$

demonstrating that equal amounts of positive and negative charge are contained in the space-charge region. The total width of this region is thus given by

$$
w = l_p + l_n
$$

$$
= \left[\left(\frac{2\varepsilon}{e}\right)\left(\frac{N_A + N_D}{N_A N_D}\right)(V_0 - V_a)\right]^{1/2}.
\tag{11.9}
$$

The electric field is given by differentiation of Eq. (11.5) as

$$
E = \begin{cases}
-\dfrac{eN_A}{\varepsilon}\left(l_p + x\right) & -l_p < x < 0 \\[3mm]
-\dfrac{eN_D}{\varepsilon}\left(l_n - x\right) & 0 < x < l_n.
\end{cases}
\tag{11.10}
$$

As shown in Fig. 11.2, the maximum value of the field occurs at $x = 0$, and its value is given by

$$
E_{max} = \frac{-2(V_0 - V_a)}{w}.
\tag{11.11}
$$

Since charge $Q = eN_a l_p$ is stored per unit area of junction in the *p*-type

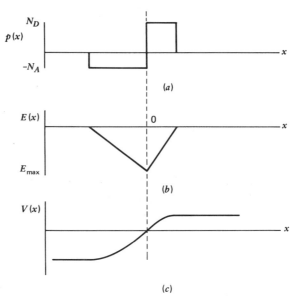

Figure 11.2. Charge density p, electric field E, and electrostatic potential V are shown as a function of position for an abrupt pn junction.

region, and charge $-Q$ is stored in the n-type region, a p-n junction can be characterized by a junction capacitance. The differential capacitance of a junction of area A is given by

$$C = A\frac{dQ}{dV} = AeN_A\frac{dl_p}{dV}$$

$$= A\left[\left(\frac{\varepsilon e}{2}\right)\left(\frac{N_A N_D}{N_A + N_D}\right)\left(\frac{1}{V_0 - V_a}\right)\right]^{1/2}$$

$$= \frac{\varepsilon A}{w}. \tag{11.12}$$

Since junction capacitance can lead to a decreased frequency response, it is important to note that the capacitance is decreased by the application of a negative bias.

The current flowing through the p-n junction can be calculated by summing the four contributions shown in Fig. 11.3. The two contributions due to negatively charged carriers (electrons) are labeled i_{nd} for the diffusion current and i_{ng} for the generation current. The diffusion current i_{nd} is made up of those electrons in the conduction band of the n-type material that diffuse into the junction with sufficient energy to surmount the potential barrier separating the *n*- and *p*-type regions. Since the height of this potential barrier decreases with increasing applied voltage V_a, this current is given by

$$i_{nd} = i_{nd,0} e^{eV_a/kT}, \tag{11.13}$$

where $i_{nd,0}$ denotes the electron diffusion current in the absence of an applied voltage. The other contribution due to electrons is the generation current i_{ng}. It results from those few electrons within the *p*-type material that are thermally excited from the valence band to the conduction band. If these thermally generated electrons encounter the junction, they are pulled into the *n*-type region, independent of the existence of an applied voltage V_a. There are analogous contributions to the current due to the motion of holes. The hole diffusion current is given by

$$i_{pd} = i_{pd,0} e^{eV_a/kT}, \tag{11.14}$$

and the hole generation current i_{pg} is independent of V_a. The total current flowing though the junction is given by the sum of these contributions as

$$i = i_{pd} + i_{nd} - i_{pg} - i_{ng}$$

$$= \left(i_{pd,0} + i_{nd,0}\right) e^{eV_a/kT} - \left(i_{pg} + i_{ng}\right). \tag{11.15}$$

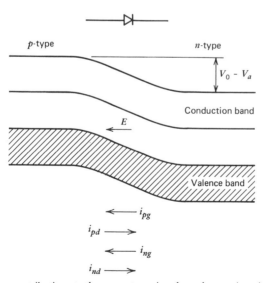

Figure 11.3. Four contributions to the current passing through a *p-n* junction are shown. The potential-barrier height is $V_0 - V_a$.

By the condition of equilibrium, no current can flow through the junction when the applied voltage is zero, and thus the two terms in parentheses must be equal. Defining the saturation current by

$$i_{sat} = i_{pg} + i_{ng}$$

$$= i_{pd,0} + i_{nd,0},$$
(11.16)

Eq. (11.15) can be reexpressed as

$$i = i_{sat}(e^{eV_a/kT} - 1).$$
(11.17)

The value of i_{sat} depends on the area of the junction and on the carrier mobilities and recombination rates. It is typically 10^{-7} to 10^{-9} A for silicon photodiodes.

The current-voltage relation given by Eq. (11.17) is modified if photoexcited carriers are present in the space-charge region; this change in the electrical characteristics induced by the presence of radiation forms the basis of the use of the p-n junction as a radiation detector. If the photon energy $h\nu$ of the incident light is greater than the band-gap energy E_g, the photon can be absorbed, creating an electron-hole pair. Figure 11.4 shows a junction that has been constructed so that the incident light is absorbed within the depletion layer. The electron and hole thus created are accelerated in opposite directions by the built-in electric field, and give rise to a

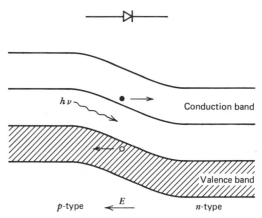

Figure 11.4. A photon of energy $h\nu$ is absorbed within the depletion layer, creating an electron-hole pair. The electron and hole are accelerated in opposite directions by the built-in electric field.

current flowing to the left in Fig. 11.4. Since by convention this constitutes a current flow in a negative sense, the current-voltage relation becomes

$$i = \frac{-\eta e P}{h\nu} + i_{\text{sat}}\left(e^{eV_a/kT} - 1\right), \tag{11.18}$$

where P is the power falling onto the detector and where η is the quantum efficiency. Equation (11.18) is displayed graphically in Fig. 11.5 for several values of the applied power P.

11.2 DETECTION USING THE PHOTODIODE

The modification of the current-voltage relation of a p-n junction diode by the presence of light falling onto the diode can be used in several distinct ways to detect radiation. In the *photoconductive mode*, the detector is biased at a negative voltage and the current passing through the detector is measured, as shown symbolically in Fig. 11.6a. Used in this manner, the detector behaves in many ways like a photoconductive detector, although typically photodiodes have a faster response than photoconductive detectors do. In addition, the rms background noise is $\sqrt{2}$ times smaller for a photodiode used in the photoconductive mode than for a true photoconductive detector, since the former shows shot noise whereas the latter shows gr noise.

Photodiodes can also be used in the *photovoltaic mode*, in which no bias voltage is applied. The short-circuit current (as in Fig. 11.6b) or the open-circuit voltage (as in Fig. 11.6c) can then be measured; or, more

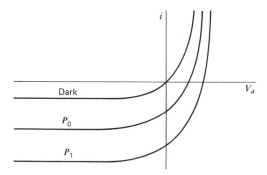

Figure 11.5. Current-voltage relation for a p-n junction photodiode in the dark and exposed to two different levels of incident power, with $P_1 > P_0$.

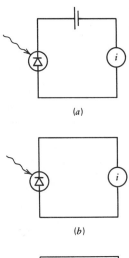

(a)

(b)

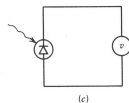

Figure 11.6. Use of a semiconductor photodiode in (a) the photoconductive mode and (b) and (c) two forms of the photovoltaic mode. The symbols i and v label ideal ammeters and voltmeters, respectively.

(c)

generally, the output voltage or current into some specified load can be measured.

The photocurrent measured at either zero voltage or at a fixed bias voltage depends linearly on the optical power, and the responsivity is given by Eq. (11.18) as

$$\mathcal{R}_i = \eta e/h\nu. \tag{11.19}$$

For $\eta = 1$, the current responsivity is of the order of 1 A/W at visible wavelengths. However, due to the exponential voltage dependence in Eq. (11.18), the open-circuit voltage depends nonlinearly on the applied power and is given by

$$V = \frac{kT}{e} \ln\left(1 + \frac{\eta eP}{h\nu i_{\text{sat}}}\right). \tag{11.20}$$

Since it is usually desirable that a radiation detector display a linear response, it is conventional to use a semiconductor photodiode as a current source when using the photovoltaic mode. One simple method for ensuring that no voltage is developed across the photodiode is to measure the

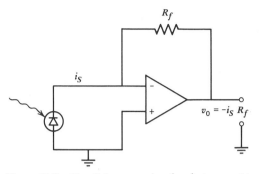

Figure 11.7. Circuit for measuring the photocurrent i_S.

Table 11.1. Cutoff Wavelengths for Several Photovoltaic Detectors

Detector Type	Cutoff Wavelength
Si	$1.1\,\mu m$
Ge	1.8
InAs	3.0
InSb	5.5

photocurrent using an operational amplifier, as shown in Fig. 11.7. Semiconductor photodiodes have been constructed using a number of different semiconductor materials, of which the most commonly encountered are Si, Ge, InAs, and InSb. The long-wavelength cutoff wavelengths of detectors using these materials are given in Table 11.1. It has been shown by Geist [J. Geist, *Appl. Opt.*, **18**, 760 (1979); see also J. Geist and E. F. Zalewski, *Appl. Phys. Lett*, **35**, 503 (1979)] that if correction is made for the reflection loss at the surface of a silicon photodiode, the quantum efficiency of such a device is equal to unity to within a few tenths of one percent for visible radiation. In addition, it has been shown that the exact value of the quantum efficiency can be determined to an accuracy sufficient to permit the use of a silicon detector as an absolute radiometric standard.

11.3 NOISE MECHANISMS

The intrinsic noise mechanism of a photovoltaic detector is shot noise in the current passing through the diode. In addition to the photocurrent $\eta eP/h\nu$ passing through the diode, each of the contributions shown in Fig.

11.3 constitutes a statistically independent contribution to the current. The mean-square current shot noise is thus given, for frequencies much less than the detector cutoff frequency, by

$$\overline{i_N^2} = 2e\left(i_{pg} + i_{pd} + i_{ng} + i_{nd} + \frac{\eta e P}{h\nu}\right)\Delta f. \tag{11.21}$$

This result can be rewritten using expressions (11.13) and (11.14) for i_{nd} and i_{pd} and the definition (11.16) for i_{sat} as

$$\overline{i_N^2} = 2e\left[i_{sat}(1 + e^{eV_a/kT}) + \frac{\eta e P}{h\nu}\right]\Delta f. \tag{11.22}$$

This result predicts that the shot noise is decreased by operating the diode under a reverse bias, since the diffusion-current contribution to the diode current is reduced by the increased height of the potential barrier separating the p- and n-type materials. However, this improved performance under reverse bias is quite difficult to achieve in practice. In real devices, the current noise often increases under a reverse bias, due presumably to $1/f$ noise in the leakage current ($1/f$ noise is discussed briefly in Section 10.3). For this reason, photovoltaic detectors used in low-noise circuits are often operated at or near zero bias voltage to minimize the leakage current.

For the case of zero applied bias voltage and zero applied optical power, the mean-square noise is given by

$$\overline{i_N^2} = 4ei_{sat}\,\Delta f, \tag{11.23}$$

which can be considered to be a form of Johnson noise, since the resistance of the junction at zero bias is given using Eq. (11.17) as

$$\frac{1}{R} = \left(\frac{di}{dV_a}\right)_{V_a=0} = \frac{ei_{sat}}{kT}, \tag{11.24}$$

implying a Johnson-noise current of

$$\overline{i_N^2} = \frac{4kT\Delta f}{R} = 4ei_{sat}\,\Delta f. \tag{11.25}$$

The contribution from Johnson noise in any load resistor used with the detector must also be included in the calculation of the total output noise.

Under conditions where sufficiently large optical power falls onto the detector, shot noise in the photocurrent can become larger than all other sources of noise. The photodiode then approximates an ideal photon detector. The resulting limitation on the NEP is discussed in Section 8.2 and, from a more general point of view, in Chapter 14.

11.4   RELATED PHOTOVOLTAIC DETECTORS

In addition to the *p-n* junction photovoltaic detector, several related types of detectors, exist.

Avalanche Photodiode

When operated at a sufficiently large negative bias voltage, photoexcited carriers within the space-charge region of a junction between fairly heavily doped *p*- and *n*-type materials are accelerated to energies sufficiently large that inelastic scattering can lead to the creation of electron-hole pairs. These free carriers are also accelerated to high energies, leading to a large current amplification by this avalanching effect. Gains of 1000 have been obtained by this process.

p-i-n Photodiode

A *p-i-n* photodiode is comprised of *p*- and *n*-type regions separated by an intrinsic region. These devices are usually designed so that most of the incident radiation is absorbed in the intrinsic region, ensuring that all the photoexcited carriers respond to the built-in electric field. In addition, since the positive and negative space-charge regions are separated by the intrinsic layer, the capacitance of such a device is smaller than that of a *p-n* junction photodiode.

Schottky Barrier Photodiode

The interface between a metal and an insulator often displays rectification due to the presence of a contact potential between the two materials. The capacitance of such devices is often very small, leading to fast time response.

BIBLIOGRAPHY

R. Dalven, *Introduction to Applied Solid State Physics*, Plenum, New York, 1980.

R. H. Kingston, *Detection of Optical and Infrared Radiation*, Springer, New York, 1978.

D. Long, in *Optical and Infrared Detectors*, R. J. Keyes, ed., Springer, New York, 1977.

S. M. Sze, *Physics of Semiconductor Devices*, Wiley-Interscience, New York, 1969.

A. Yariv, *Introduction to Optical Electronics*, Holt, Rinehart, and Winston, New York, 1971.

PROBLEM

1   A photovoltaic detector having dynamic resistance R at zero bias voltage
is operated at temperature T. It is subjected to a background photon
flux equal to $\Phi_p = E_p A$, where E_p is the photon irradiance of the
background and A is the detector area. What conditions must E_p, A, R,
and T satisfy in order for the detector to be background-noise limited?

Calculate the value of E_p for a detector with a 10-μm wavelength
cutoff exposed to a 300-K background within a 60° (full angle) field of
view, and determine numerically the requirements for R and A assuming that the detector is operating at 77 K.

12

Coherent Detection

The detection systems that have been considered in the previous several chapters utilized what might be called direct detection. In direct detection, the output electrical signal is a current (or voltage) that ideally is linearly proportional to the signal power. As such, no information regarding the phase of the optical signal is retained in the electrical output.

A heterodyne detection system, in contrast, produces an output current (or voltage) that ideally is proportional to the electric field strength of the optical signal. The phase of the optical field is thus preserved in the phase of the electrical signal, and in this sense the detection process can be considered to be coherent. Heterodyne detection can usefully be employed whenever it is necessary to preserve the phase of the signal field, or when extremely good frequency resolution is required. The heterodyne detection process is, however, intrinsically quite noisy, and this places a rather serious constraint on the ultimate sensitivity of a heterodyne detection system operating at optical frequencies.

12.1 SQUARE-LAW DETECTORS

Most radio-frequency receivers utilize heterodyne detection. Therefore, it is common to use the terminology relevant to radio-frequency heterodyne receivers to describe heterodyne receivers operating even at optical frequencies. All heterodyne detection systems operate by mixing the field to be detected with a single-frequency local oscillator field in a nonlinear circuit element, which ideally is a square-law detector. In radio-frequency systems, a diode is often used as the square-law detector. The current-voltage characteristic of a typical diode is shown in Fig. 12.1. A power-series expansion of v in terms of i performed at $i = 0$ will have the form

$$v = Ji + Ki^2 + Li^3 + \cdots .\tag{12.1}$$

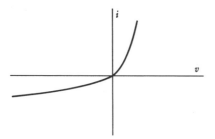

Figure 12.1. Characteristic curve of a typical diode.

The voltage thus has a contribution proportional to i^2, and in this sense the diode can be referred to as a square-law detector. Since the power contained in a signal $i(t)$ is proportional to $i^2(t)$, the voltage generated by the diode when exposed to the signal $i(t)$ will have a contribution proportional to the incident power.

The optical analog of the square-law detector is the photon detector, which for simplicity we consider to be a photodiode. The photocurrent produced when instantaneous power $P(t)$ falls onto the detector is given by

$$i(t) = \frac{\eta e P(t)}{h\nu}. \qquad (12.2)$$

Since the optical power $P(t)$ is proportional to the square of the electric field strength $E(t)$, the photon detector behaves as a square-law detector.

12.2 OPTICAL HETERODYNE DETECTION

The optical heterodyne detection process is illustrated schematically in Fig. 12.2. A highly monochromatic laser field of frequency ω_L serves as a local oscillator. It is combined with a signal field of frequency ω_S through the use of a beam splitter, and the combined field is allowed to fall onto a

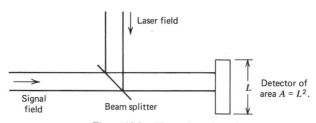

Figure 12.2. Heterodyne detection.

photon detector, often known in this context as a photomixer. It is assumed that the two optical beams are well collimated and are aligned so that their wave fronts are parallel. The field at the detector can then be taken as the real part of the expression

$$E(t) = E_L e^{-i\omega_L t} + E_S e^{-i\omega_S t}, \qquad (12.3)$$

where the field strengths are measured after the beam splitter and where it has been assumed for simplicity that the signal and laser fields are linearly polarized in the same direction. Using Eqs. (1.11) and (1.23), the instantaneous value of the power falling onto the detector is given as

$$P(t) = \sqrt{\frac{\varepsilon}{\mu}} A [\operatorname{Re} E(t)]^2. \qquad (12.4)$$

When Eq. (12.3) is inserted into this expression and the result is averaged over many optical periods, one obtains an expression for the cycle-averaged optical power given by the real part of

$$P(t) = \frac{1}{2} \sqrt{\frac{\varepsilon}{\mu}} A [|E_L|^2 + |E_S|^2 + E_S E_L^* e^{-i(\omega_S - \omega_L)t}]. \qquad (12.5)$$

An explicit expression for the photocurrent can now be obtained by using this form for $P(t)$ in Eq. (12.2) for $i(t)$. This expression is simplified through use of the notation

$$i_L \equiv \frac{\eta e A}{2h\nu} \left(\frac{\varepsilon}{\mu}\right)^{1/2} |E_L|^2, \qquad (12.6)$$

$$i_S \equiv \frac{\eta e A}{2h\nu} \left(\frac{\varepsilon}{\mu}\right)^{1/2} |E_S|^2. \qquad (12.7)$$

These expressions give the respective values of the photocurrent if only the laser field or only the signal field is present. The general form for $i(t)$ is then

$$i(t) = i_L + i_S + 2\sqrt{i_L i_S} \cos [(\omega_S - \omega_L)t + \phi], \qquad (12.8)$$

where ϕ represents the relative phase between E_S and E_L, and is given by

$$\phi = \arctan\left(\frac{\operatorname{Im} E_L}{\operatorname{Re} E_L}\right) - \arctan\left(\frac{\operatorname{Im} E_S}{\operatorname{Re} E_S}\right). \qquad (12.9)$$

The photocurrent thus has a dc component and a component oscillating at the beat frequency $\omega_{i.f.} = \omega_S - \omega_L$, often known as the intermediate

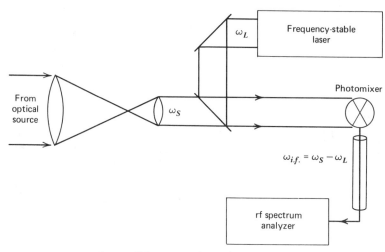

Figure 12.3.   Optical heterodyne receiver.

frequency. The heterodyne signal is contained entirely in this frequency component. The strength of the heterodyne signal can thus be characterized by the mean-square value of this frequency component of the current, that is, by

$$\overline{i^2_{\text{i.f.}}} = 2i_L i_S. \qquad (12.10)$$

Figure 12.3 shows symbolically the layout of an optical heterodyne receiver. It is important that the laser frequency be highly stable since the beat frequency $\omega_{\text{i.f.}}$ will track any fluctuations in laser frequency. In practice, the optical signal to be detected is often spread over a broad frequency band. An optical frequency component ω_S will give rise to an i.f. component in the photocurrent only if the beat frequency $\omega_{\text{i.f.}} = \omega_S - \omega_L$ lies within range covered by the frequency response of the photomixer. Typically, this frequency response is limited to frequencies less then approximately 1 GHz. The optical frequency bandwidth $\Delta\omega_S$ that can be detected is thus equal to the i.f. bandwidth $\Delta\omega_{\text{i.f.}}$, which is limited by the frequency response of the photomixer. The i.f. signal can be further processed by sending it to a radio-frequency spectrum analyzer, as shown in the figure.

12.3 SINGLE MODE RECEPTION

An optical heterodyne receiver is often called a single-mode receiver in the sense that it can collect signal from only one transverse mode of the radiation field. This section will examine this aspect of the heterodyne reception process and will formulate precisely what is meant by the mode structure of a radiation field in a free space.

Our initial discussion of heterodyne detection assumed that the signal and laser wave fronts were perfectly parallel at the detector. Figure 12.4 shows the case in which the signal wave front is inclined at the angle θ to the laser wavefront. When $\theta = 0$, the difference-frequency signals from all points on the surface of the detector contribute with the same phase to the i.f. signal. For $\theta \neq 0$, different points on the surface will contribute i.f. signals with different phases, causing the total signal to decrease. The heterodyne signal will drop to zero when

$$\theta L = \lambda,$$

λ being the wavelength of the signal wave. Since the solid angle Ω from which the receiver can collect signal must be of order θ^2, we deduce the relation

$$A\Omega \approx \lambda^2.$$

An exact calculation of the effective solid angle from which a heterodyne receiver of active area A can respond to incident radiation can be performed by integrating the angular response of the receiver over all angles. Such a calculation (see, e.g., the work by Kingston cited at the end of this chapter) shows that A is related to the projected solid angle by

$$A\Omega_{proj} = \lambda^2. \tag{12.11}$$

[The projected solid angle was defined by Eq. (4.47). It differs from the usual solid angle by a factor of the order of $\cos \theta$.] The quantity $A\Omega_{proj}$ was called the *étendue* in Chapter 4. This quantity was shown to be conserved for any optical system, and thus Eq. (12.11) can equally well be used to

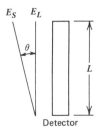

E_S E_L

θ

L

Detector

Figure 12.4. Illustration of signal and laser wave fronts at the photomixer.

calculate the projected acceptance solid angle at any reference surface of the optical system.

The relation expressed by Eq. (12.11) can be said to define a single transverse mode of the radiation field. In order to establish this connection, it is necessary to generalize the definition of a field mode that was used previously. In the discussion of blackbody radiation in Chapter 3, the modes of the radiation field were defined as the standing wave solutions to the wave equation for boundary conditions appropriate to an enclosure with perfectly (electrically) conducting walls. For that situation, the size and shape of the enclosure were entirely arbitrary, so long as its dimensions were much larger than an optical wavelength, since only the density of modes appeared in any of the results.

In problems involving the propagation and detection of radiation, it is more natural to define the field modes in terms of traveling-wave solutions to the wave equation, since such solutions can allow for a flow of energy. It is still convenient to define the field modes in terms of a reference volume, but now the shape of this volume is much less arbitrary, since the surface of the detector serves as a natural reference surface with which to define field modes. Assuming for convenience that the detector is a rectangle with sides L_x and L_y, an appropriate reference volume is shown in Fig. 12.5. The longitudinal extent L_z of the cavity is still arbitrary. Thus we wish to consider solutions to the wave equation in the form of traveling waves, that is, waves of the form

$$\mathbf{E}(\mathbf{r}, t) = \mathbf{E}_\alpha(\mathbf{k}) e^{i(\mathbf{k} \cdot \mathbf{r} - \omega t)}, \qquad (12.12a)$$

with

$$\mathbf{k} \cdot \mathbf{E}_\alpha(\mathbf{k}) = 0, \qquad (12.12b)$$

where $\alpha = 1, 2$ is an index to account for the two allowed directions of polarization. By analogy to the treatment of Chapter 3 in terms of standing waves, we wish to treat the present situation as a boundary-value problem. The simplest boundary conditions that ensure that the mode density for traveling-wave solutions is the same as that for standing-wave solutions are the so-called periodic boundary conditions:

$$\mathbf{E}(0, y, z, t) = \mathbf{E}(L_x, y, z, t), \qquad (12.13a)$$

$$\mathbf{E}(x, 0, z, t) = \mathbf{E}(x, L_y, z, t), \qquad (12.13b)$$

$$\mathbf{E}(x, y, 0, t) = \mathbf{E}(x, y, L_z, t). \qquad (12.13c)$$

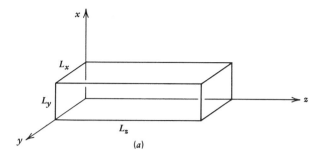

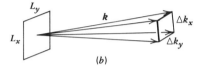

Figure 12.5. (*a*) Reference volume for calculating traveling-wave field modes. (*b*) Four adjacent transverse field modes define a solid angle.

These boundary conditions ensure that the field will be periodic in x with period L_x, and similarly for y and z. They are thus the appropriate boundary conditions to apply to an enumeration of the traveling-wave solutions in a finite region of space, since each field mode is completely defined by its behavior in only that region of space corresponding to the reference volume. By applying the boundary conditions (12.13) to the solutions (12.12), the allowed values of the wave vector $\mathbf{k} = (k_x, k_y, k_z)$ are seen to be of the form

$$k_x = \frac{2\pi}{L_x} n_x,$$

$$k_y = \frac{2\pi}{L_y} n_y,$$

$$k_z = \frac{2\pi}{L_z} n_z,$$

where n_x, n_y, and n_z can be any positive or negative integer or zero.

It is natural to treat the allowed values of k_x and k_y separately from the allowed values of k_z, since the former depend on physically meaningful dimensions L_x and L_y. By way of nomenclature, we can say that the allowed values of k_x and k_y define the transverse modes of the field. Figure 12.5b shows four adjacent allowed values of k_x and k_y. Together, they define a projected solid angle that may be associated with a single transverse field mode, which is given by

$$\Omega_{\text{proj}} = \frac{\Delta k_x \Delta k_y}{k^2} = \frac{\lambda^2}{L_x L_y},$$

or, letting $A = L_x L_y$, by

$$\Omega_{\text{proj}} = \lambda^2/A. \tag{12.14}$$

This result can be interpreted physically as a statement that no source of projected solid angle greater than λ^2/A can provide spatially coherent illumination over a region of area A. For a source of projected solid angle Ω'_{proj} larger than that given by Eq. (12.14), this equation implies that one can completely specify the field distribution across a detector of aperture $A = L_x L_y$ by specifying the field amplitudes $\mathbf{E}_\alpha(\mathbf{k})$ of a finite number $\sim (A\Omega'_{\text{proj}}/\lambda^2)$ of field modes.

The allowed values of k_z can likewise be associated with the longitudinal modes of the radiation field. Since the allowed values of k_z are spaced by $2\pi/L_z$, the number of modes with values of k_z in the interval dk_z is equal to $dk_z L_z/2\pi$. Thus the number of modes per unit length in the interval dk_z is equal to $dk_z/2\pi$.

The mode density per unit frequency interval is now obtained by first noting that the relation between $\mathbf{k}$ and ω is of the form

$$k_x^2 + k_y^2 + k_z^2 = \frac{\omega^2}{c^2}. \tag{12.15}$$

For a given transverse mode, k_x and k_y are fixed, and thus dk_z and $d\omega$ must vary as

$$k_z \, dk_z = \frac{\omega \, d\omega}{c^2}. \tag{12.16}$$

Since the quantity $k_z c/\omega$ is equal to $\cos \theta$, where θ is the angle between $\mathbf{k}$ and the z-axis, this result can be expressed as

$$dk_z = \frac{d\omega}{c \cos \theta}. \tag{12.17}$$

The density $\rho_{1d}(\omega)$ of longitudinal modes per unit length and per unit angular frequency interval is thus given by

$$\rho_{1d}(\omega) = (2\pi c \cos\theta)^{-1}, \tag{12.18}$$

and thus the total number of such modes per unit length in the frequency interval $d\omega$ is equal to $\rho_{1d}(\omega)\,d\omega$. Since these modes represent traveling waves, they will pass through the plane $z = 0$ defining the end of the reference volume at a rate r given by the product of $\rho_{1d}(\omega)\,d\omega$ with the z component of the phase velocity, $c\cos\theta$, giving

$$r = \frac{d\omega}{2\pi} = d\nu. \tag{12.19}$$

Thus the rate at which a detector receives longitudinal field modes from a single transverse field mode is numerically equal to the frequency breadth of the field. In a time interval T, therefore, a number $\Delta\nu T$ modes of the field fall onto the detector from a single transverse field mode. This number $\Delta\nu T$ can be interpreted more physically as the maximum number of independent measurements of the field amplitude that can be performed in this time interval.

These results can be illustrated more explicitly by considering a detector of area A collecting radiation in a narrow spectral interval $\Delta\nu$ from a blackbody source at temperature T that subtends a projected solid angle Ω'_{proj}. The total power collected by the detector is thus given by

$$P = L_\nu(T) A \Omega'_{\text{proj}} \Delta\nu, \tag{12.20}$$

where $L_\nu(T)$ is the Planck radiation function

$$L_\nu(T) = \frac{2h\nu^3/c^2}{e^{h\nu/kT} - 1}. \tag{12.21}$$

The power falling onto the detector can alternatively be calculated by arguing that each field mode in the interval $\Delta\nu$ contains on the average $\bar{n} = (e^{h\nu/kT} - 1)^{-1}$ photons of energy $h\nu$, and that these modes fall onto the detector at some fixed rate r. By equating the two resulting expressions for the incident power, the rate r is seen to be given by

$$r = 2\left(\frac{A\Omega'_{\text{proj}}}{\lambda^2}\right)\Delta\nu, \tag{12.22}$$

which is in agreement with the preceding results since both field polarizations have been included in the present calculation, giving rise to the factor of 2, and since the quantity in parentheses is the number of transverse field modes interacting with the detector.

12.4 NOISE IN HETERODYNE DETECTION

The predominant noise source in a well-designed optical heterodyne detection system is shot noise in the photocurrent. Under the condition $i_L \gg i_S$, which requires that the laser power be much greater than the signal power, the mean-square current noise is given by Eq. (8.42) as

$$\overline{i_N^2} = 2ei_L\,\Delta f_{\text{i.f.}}, \tag{12.23}$$

where $\Delta f_{\text{i.f.}}$ is the electrical bandwidth of the i.f. channel, which, as mentioned earlier, is equal to the optical bandwidth of the detected signal radiation and is usually limited primarily by the frequency response of the photomixer. Since the mean-square i.f. current is given by Eq. (12.10) as

$$\overline{i_{\text{i.f.}}^2} = 2i_L i_S,$$

the square of the current signal-to-noise ratio in the i.f. channel is given by

$$\left(\frac{S}{N}\right)_{\text{i.f.}}^2 \equiv \frac{\overline{i_{\text{i.f.}}^2}}{\overline{i_N^2}} = \frac{i_S}{e\,\Delta f_{\text{i.f.}}}$$

$$= \frac{\eta P_S}{h\nu\,\Delta f_{\text{i.f.}}}, \tag{12.24}$$

where the last form was obtained by using the standard relation (12.2) between optical power and photocurrent. Since this result does not depend on the value of i_L, it is possible to increase the laser power until the shot noise in the photocurrent swamps all other noise sources, justifying the use of Eq. (12.23) for the total current noise. Equation (12.24) can be considered to give the intrinsic signal-to-noise ratio for heterodyne detection since, in general, the i.f. bandwidth $\Delta f_{\text{i.f.}}$ cannot be changed without also changing the signal power P_S. If, for instance, the optical signal power per unit frequency interval P_ν is constant across the bandwidth of the receiver, the signal power can be represented by

$$P_S = P_\nu\,\Delta f_{\text{i.f.}}, \tag{12.25}$$

and Eq. (12.24) becomes

$$\left(\frac{S}{N}\right)^2_{\text{i.f.}} = \frac{\eta P_\nu}{h\nu}.\tag{12.26}$$

Thus, under the stated conditions, the i.f. signal-to-noise ratio is independent of the i.f. bandwidth and depends only on the signal power per unit frequency interval. If the signal field is provided by a blackbody source of temperature T, the quantity P_ν is given by

$$P_\nu = \tfrac{1}{2}L_\nu A\Omega_{\text{proj}},\tag{12.27}$$

where the factor of $\tfrac{1}{2}$ accounts for the single polarization detected by a heterodyne receiver. Using Eq. (3.69) for L_ν and the relation $A\Omega_{\text{proj}} = \lambda^2$, this result can be expressed as

$$P_\nu = \frac{h\nu}{e^{h\nu/kT} - 1} = \bar{n}h\nu,\tag{12.28}$$

with $\bar{n}$ denoting as before the average number of photons per mode. The i.f. signal-to-noise ratio (12.26) thus becomes

$$\left(\frac{S}{N}\right)^2_{\text{i.f.}} = \eta\bar{n}.\tag{12.29}$$

If the quantum efficiency has its maximum possible value $\eta = 1$, a signal-to-noise ratio of unity requires that $\bar{n} = 1$, which required a blackbody temperature given by

$$T_N = \frac{h\nu}{k\ln 2}.\tag{12.30}$$

This temperature can be considered to be the noise temperature of a heterodyne receiver, since a source of this temperature is required to produce a signal-to-noise ratio of unity. The noise temperature increases linearly with the optical frequency ν. At a frequency corresponding to a wavelength of 1 μm, the noise temperature equals 21,000 K, illustrating that optical heterodyne reception is in fact quite noisy.

12.5 POST-DETECTION FILTERING

The discussion of the previous section showed that the signal-to-noise ratio in the i.f. channel of an optical heterodyne receiver is often quite low.

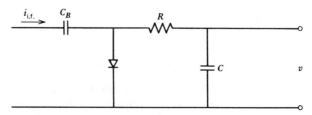

Figure 12.6.   Circuit for detecting and averaging the i.f. signal.

The signal-to-noise ratio can often be significantly improved by averaging the i.f. signal for a time interval considerably greater than the reciprocal of the i.f. bandwidth.

A circuit that can perform such an average is shown in Fig. 12.6. The blocking capacitor of value C_B eliminates the dc components from the photocurrent. The diode in this circuit is used as a square-law detector, and thus develops a voltage proportional to the mean-square i.f. current. The noise in this voltage signal is reduced by the filtering action of the RC (resistance-capacitance) low-pass filter.

In order to calculate the resulting increase in the output signal-to-noise ratio, we assume that the diode characteristic is of the form

$$v(t) = Ji(t) + Ki^2(t), \qquad (12.31)$$

where J and K are some constants. The dc component of the voltage developed across the diode is thus given using Eq. (12.10) by

$$v_0 = 2Ki_L i_S. \qquad (12.32)$$

The heterodyne signal is thus detected at this point in that the signal has been reduced to a dc level. Similarly, the shot noise in the photocurrent develops a voltage in passing through the diode that is given by

$$v_N = 2Kei_L \Delta f_{\text{i.f.}}. \qquad (12.33)$$

The noise voltage appearing at the output of the circuit will be reduced by the RC filter. A circuit with a *response* time $\tau = RC$ has an equivalent electrical bandwidth given using Eq. (7.9) as $\Delta f_0 = 1/4RC$, and by comparison with Eq. (7.11) such a bandwidth is associated with an equivalent integration time of $T = 2\tau = 2RC$. Thus, in a time interval of duration $T = 2RC$, the noise voltage can be sampled independently $2 \Delta f_{\text{i.f.}} RC$ times, leading to a reduction in the voltage noise by the square root of this number. The output voltage noise is thus given by

$$v_N = \frac{2 K e i_L \Delta f_{\text{i.f.}}}{(\Delta f_{\text{i.f.}}/2 \Delta f_0)^{1/2}} . \tag{12.34}$$

The output voltage signal-to-noise ratio is thus equal to the ratio of Eqs. (12.32) and (12.34), giving

$$\left(\frac{S}{N}\right)_0 = \frac{i_S}{e\sqrt{2 \Delta f_{\text{i.f.}} \Delta f_0}} . \tag{12.35}$$

For the case of an optical signal derived from a radiation field of average occupation number $\bar{n}$, the arguments leading from Eqs. (12.24) to (12.29) can be used to express this result in the form

$$\left(\frac{S}{N}\right)_0 = \eta \bar{n} \sqrt{\frac{\Delta f_{\text{i.f.}}}{2 \Delta f_0}} . \tag{12.36}$$

Thus the output signal-to-noise ratio has been increased by the square root of the ratio of the i.f. bandwidth to twice the *post-detection* bandwidth Δf_0.

An exact comparison between the sensitivity of a heterodyne detection system and a direct detection system is complicated by the fact that the heterodyne system can detect only one transverse mode of the radiation field and thus responds to the radiance of the optical field, whereas a direct detector responds to the total power incident on the detector. With this caution in mind, we can discuss the heterodyne detection process using the descriptors usually applied to direct detectors. Equation (12.35) for the signal-to-noise ratio can be expressed in terms of the signal power giving

$$\left(\frac{S}{N}\right)_0 = \frac{\eta P_S}{h\nu (2 \Delta f_{\text{i.f.}} \Delta f_0)^{1/2}} . \tag{12.37}$$

The NEP is obtained by solving for P_S under the condition $(S/N)_0 = 1$, giving

$$P_N = \frac{h\nu \sqrt{2 \Delta f_{\text{i.f.}} \Delta f_0}}{\eta .} \tag{12.38}$$

The NEP of a direct detector in the presence of background power P_B was shown in Eq. (8.60) to be of the form

$$P_N = \sqrt{\frac{2h\nu P_B \Delta f_0}{\eta}} \, .$$

Comparison of these expressions shows that the intrinsic noise in heterodyne detection corresponds to an effective background power given by

$$(P_B)_{\text{eff}} = \frac{h\nu \, \Delta f_{\text{i.f.}}}{\eta} \, . \tag{12.39}$$

For the case $\eta = 1$, this effective background is equivalent to the contribution of one photon per mode. The source of this noise can be understood in two different ways. The point of view taken up till now has been that noise originates as shot noise in the photocurrent. We have obtained the perhaps surprising result, however, that (for $i_L \gg i_S$) the NEP of a heterodyne detection system is independent of the magnitude of the shot noise in the photocurrent, although the NEP is limited by an effective background that is equivalent to $1/\eta$ photons per mode. This occurrence suggests that noise in the heterodyne detection process has a more fundamental origin than the one given above, namely that it constitutes a form of quantum noise. This alternative point of view is presented in the following paragraph.

In microscopic terms, the heterodyne detection process can be viewed as the interaction of signal and laser photons in a nonlinear element to create an i.f. photon. Assuming for illustration that $\omega_L > \omega_S$, the process can be viewed, as in Fig. 12.7, as stimulated two-photon emission from the virtual level created by the laser photon at ω_L. The rate at which this process proceeds can be predicted by the rules of quantum field theory to be of the form

$$r = Mn_L(n_S + 1), \tag{12.40}$$

where M is a constant that depends on the strength of the optical nonlinearity and where n_L and n_S denote the average number of detected laser and signal photons per mode. The absence of an applied signal field implies that $n_S = 0$, but the generation of i.f. photons can still proceed by spontaneous emission at a rate equal to Mn_L. Spontaneously emitted i.f. photons constitute from this microscopic point of view the source of noise in the heterodyne detection process. Since the stimulated and spontaneous rates become equal for $n_S = 1$, the background present in a heterodyne receiver corresponds to one detected photon per mode; and since a heterodyne

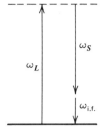

Figure 12.7. Microscopic description of heterodyne detection.

receiver can detect only a single transverse mode of the radiation field, the detected power in this background is given by

$$P = h\nu \Delta f_{\text{i.f.}}. \tag{12.41}$$

Since only a fraction η of the incident radiation is detected, the effective background power is given by

$$(P_B)_{\text{eff}} = \frac{h\nu \Delta f_{\text{i.f.}}}{\eta}, \tag{12.42}$$

in agreement with the result (12.39) based on a consideration of the shot noise in the photocurrent.

BIBLIOGRAPHY

R. H. Kingston, *Detection of Optical and Infrared Radiation*, Springer, Berlin, 1978, Chapter 3.

M. Ross, *Laser Receivers*, Wiley, New York, 1966.

M. C. Teich, *Proc. IEEE*, **54**, 1350 (1968); also in *Semiconductors and Semimetals*, Vol. 5, R. K. Willardson and A. C. Beer, eds., Academic, New York, 1970.

PROBLEMS

1 Determine the signal-to-noise ratio that can be achieved with a 1-Hz post-detection bandwidth in the detection of the radiation from a tungsten lamp using a laser heterodyne receiver. You may assume that the laser wavelength is 5000 Å, that the lamp temperature is 2800 K and that the photomixer is an ideal, unit-quantum-efficiency, photon detector whose (exponential) response time is 5 ns.

2   The output of a laser is divided into two beams, one of which serves as
the local oscillator of a laser heterodyne receiver. The other beam is
reflected at near normal incidence from a mirror whose position is
modulated, in a direction perpendicular to its surface, as

$$z = z_1 \cos \omega_1 t,$$

and it is then detected by the receiver. Determine the form of the current
$i(t)$ produced by the receiver and describe to what extent noise will
affect a measurement of $i(t)$.

13

Thermal Detectors

The analysis of thermal detectors is in many ways more complicated than that of photon detectors. Thermal detectors operate by monitoring the temperature change induced in an active element by the absorption of radiation. Kirchhoff's law requires, however, that the emissivity of this element must equal its absorptivity; thus a thermal detector must also be an emitter of thermal radiation. The fundamental limitation to the sensitivity of a thermal detector is set by fluctuations in the temperature of the active element. Fluctuations in the rates at which photons are both absorbed and emitted by the element contribute to these temperature fluctuations, and thus an analysis of noise processes in thermal detectors is intrinsically linked to an understanding of fluctuations in the radiation field.

The present chapter presents a general discussion of thermal detection and an analysis of two of the most commonly used thermal detectors, the bolometer and the pyroelectric detector. An elementary treatment of noise, based on fluctuations in thermal equilibrium, is presented in this chapter. A more detailed treatment, based on an analysis of fluctuations in a thermal radiation field, is presented in Chapter 14.

13.1 ELEMENTARY THEORY OF THERMAL DETECTORS

Thermal detectors operate by the temperature rise induced in an active element by the absorption of incident radiation. As mentioned in Section 7.1, several mechanisms for measuring the temperature of the active element are available. For the present, however, we shall ignore the details of these mechanisms and discuss only those aspects of the responsivity and sensitivity that are common to all thermal detectors. The present discussion thus deals with an *ideal thermal detector*, in that we shall assume that no inaccuracy or noise is introduced into the detection system by the mechanism used to measure the temperature change induced by the radiation field.

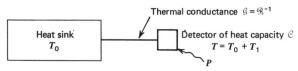

Figure 13.1. Thermal detection process.

Figure 13.1 shows schematically a thermal detection system. Radiation of power P is allowed to fall onto the detector element, causing its temperature to increase from T_0 to $T_0 + T_1$. The following discussion assumes that $T_1 \ll T_0$. The detector is assumed to have heat capacity $\mathcal{C}$ and to be connected to a heat sink of constant temperature T_0 through a link of thermal conductance $\mathcal{G}$.* The hat capacity is defined by

$$\mathcal{C} = \frac{dQ}{dT_1}, \tag{13.1}$$

where dQ is the additional heat stored in the detector when its temperature is changed by an amount dT_1. The thermal conductance $\mathcal{G}$ is defined by the relation

$$P_0 = \mathcal{G}T_1, \tag{13.2}$$

where P_0 is the heat flow from the detector to the heat sink if the detector is maintained at a constant temperature T_1 above that of the heat sink. If the power falling onto the detector is the time-varying quantity $P(t)$ the time evolution of the incremental temperature T_1 of the detector element is

---

*We are following the convention here of denoting by script letters the thermal properties that are analogous to the electrical properties denoted by the same roman letter. For instance, the heat capacity $\mathcal{C}$ is the analog of the electrical capacity C. In fact, thermal circuits are formally identical to electrical circuits through use of the following substitutions:

Temperature difference $T_1 \leftrightarrow$ voltage v

Rate of heat flow $P \leftrightarrow$ current i

Heat capacity $\mathcal{C} \leftrightarrow$ electrical capacity C

Thermal conductance $\mathcal{G} \leftrightarrow$ electrical conductance G

where the electrical conductance G is the reciprocal of the electrical resistance R.

governed by

$$\mathcal{C}\frac{dT_1}{dt} + \mathcal{G}T_1 = \varepsilon P(t), \tag{13.3}$$

where ε denotes the fraction of the incident power absorbed by the detector element (i.e., its absorptivity). Let us first consider the temperature response to a step-function input of the form

$$P(t) = \begin{cases} 0 & t < 0 \\ P_0 & t \geqslant 0. \end{cases} \tag{13.4}$$

The solution to Eq. (13.3) for this form of $P(t)$ is given by

$$T_1(t) = \begin{cases} 0 & t < 0 \\ \dfrac{\varepsilon P_0}{\mathcal{G}}(1 - e^{-t/\tau_T}), \end{cases} \tag{13.5}$$

where the thermal response time defined by

$$\tau_T = \frac{\mathcal{C}}{\mathcal{G}} \tag{13.6}$$

has been introduced. Thus the magnitude of the response is increased by making the thermal conductance $\mathcal{G}$ as small as possible. The response time can then be made short only by decreasing the heat capacity $\mathcal{C}$. These conditions require that thermal detectors be quite small physically and that they be thermally isolated from the immediate environment. Thus thermal detectors are often quite fragile.

Thermal detection systems often utilize modulated (or "chopped") radiation. It is possible to represent one harmonic component of the incident radiation as the real part of

$$P(t) = P_0 e^{i2\pi ft}. \tag{13.7}$$

The steady-state solution to Eq. (13.3) then is given by

$$T_1(t) = T_1 e^{i2\pi ft}, \tag{13.8a}$$

where

$$T_1 = \frac{\varepsilon P_0}{\mathcal{G} + i2\pi f\mathcal{C}}. \tag{13.8b}$$

The magnitude of the response is thus

$$|T_1| = \frac{\varepsilon P_0/\mathcal{G}}{\left(1 + 4\pi^2 f^2 \tau_T^2\right)^{1/2}}, \qquad (13.9)$$

leading to a thermal responsivity (whose units might be K/W) of the form

$$\mathcal{R}_T = \frac{\varepsilon/\mathcal{G}}{\left(1 + 4\pi^2 f^2 \tau_T^2\right)^{1/2}}. \qquad (13.10)$$

In the absence of applied radiation, the only noise present in an ideal thermal detector is temperature noise, that is, fluctuations in the temperature of the active element that occur even in thermal equilibrium. The mean-square magnitude of these fluctuations can be derived using the techniques of nonequilibrium statistical mechanics (see, e.g., L. D. Landau and E. M. Lifshitz, *Statistical Physics*, Addison-Wesley, Reading, Mass., 1969, p. 350). This magnitude can be represented as

$$\overline{(\Delta T_1)^2} = \frac{kT_0^2}{\mathcal{C}}. \qquad (13.11)$$

This result is sometimes referred to as the Fowler–Einstein equation. It is derived under the assumptions that the detector can exchange energy only with the heat sink, that the two are in thermal equilibrium at the temperature T_0, and that the heat capacity of the detector is much less than that of the heat sink.

We now wish to determine the spectral density of these fluctuations. This is most quickly accomplished by exploiting the formal identity between the thermal and electrical circuits noted earlier. In the discussion of Johnson noise presented in Chapter 8, the mean-square noise voltage $\overline{v_N^2}$ was given by Eq. (8.62), and the spectral density for a circuit characterized by a single decay time τ was given by Eq. (8.65) as

$$\overline{v_N^2}(f) = \frac{4\tau_T \overline{v_N^2}}{1 + (2\pi f\tau)^2}.$$

For the present case of temperature fluctuations, the mean-square temperature fluctuation $\overline{(\Delta T_1)^2}$ and its spectral density $\overline{(\Delta T_1)^2}(f)$ must be related in the same way, and thus the spectral density is given by*

$$\overline{(\Delta T_1)^2}(f) = \frac{4kT^2/\mathcal{G}}{1 + (2\pi f\tau_T)^2}.\qquad(13.12)$$

The rms temperature fluctuation in a narrow bandwidth Δf is thus given by $(\Delta T)_{rms} = [\overline{(\Delta T_1)^2}(f)\,\Delta f]^{1/2}$. The signal power required to produce a temperature increase equal to this rms temperature noise defines the noise equivalent power (NEP) and is given by

$$P_N = \frac{(\Delta T)_{rms}}{|\mathcal{R}_T(f)|},\qquad(13.13)$$

or, using Eqs. (13.10) and (13.12), as

$$P_N = \frac{1}{\varepsilon}\sqrt{\mathcal{G}4kT_0^2\,\Delta f}\quad\text{for } f \ll 1/\tau_T.\qquad(13.14)$$

The NEP is thus improved by making the emissivity of the detector larger or by making the thermal conductance or the temperature of the device smaller. The emissivity and thermal conductance are not, however, entirely independent of one another. The thermal conductance can be decreased by eliminating all conductive and convective cooling of the detector's active element, but radiative cooling cannot be eliminated since it is necessary that a detector be able to interact with the radiation field. The power leaving the detector through thermal emission is given by

$$P = A\varepsilon'\sigma T^4,\qquad(13.15)$$

where A is the detector area and ε' is its emissivity. We allow ε' to be different from ε since the spectral composition of the incident radiation can differ from that of the emitted radiation. If the detector temperature is raised by an amount dT, the power leaving the detector must increase by the

---

*This same result can be derived by a formal application of the argument presented in Chapter 8 for Johnson noise. It can also be obtained less directly but more elegantly by exploiting the relations that always exist between the fluctuations that occur in a system and the processes that can lead to the dissipation of energy. This relation is known either as the fluctuation-dissipation theorem or as the generalized Nyquist theorem, and it is discussed by H. B. Callen and T. A. Welton, *Phys. Rev.*, **87**, 471 (1952) and by F. Reif, *Fundamentals of Statistical and Thermal Physics*, McGraw-Hill, New York, 1965, Chapter 15.

amount

$$dP = 4A\varepsilon'\sigma T^3 \, dT. \tag{13.16}$$

The thermal conductance is thus given by

$$\mathcal{G} = \frac{dP}{dT} = 4A\varepsilon'\sigma T_0^3. \tag{13.17}$$

This expression thus represents the smallest allowed value of the thermal conductance of a detector of area A and emissivity ε' that is held at ambient temperature T_0. The corresponding value of the NEP is thus

$$P_N = 4\sqrt{\frac{A\sigma kT^5 \Delta f \, \varepsilon'}{\varepsilon^2}}, \tag{13.18a}$$

implying a specific detectivity given by

$$D^* = \frac{1}{4}\sqrt{\frac{\varepsilon^2}{\varepsilon'\sigma kT^5}}. \tag{13.18b}$$

If $\varepsilon = \varepsilon' = 1$ and $T = 300$ K, this expression is equal to $D^* = 1.8 \times 10^{10}$ cm Hz$^{1/2}$/W.

13.2   THE BOLOMETER

As an illustration of the general ideas presented in Section 13.1, we now analyze the operation of the bolometer, illustrated in Fig. 13.2. Radiation of power P is allowed to fall onto the bolometer element, thus changing the value of its resistance R_B. This in turn causes the output voltage v to vary. In practice, it is often convenient to choose the load resistor of value R_L to be of identical construction to the bolometer but shielded from the incident radiation, so that variations in the ambient temperature or of the bias emf V are automatically compensated in the output voltage. The magnitude of the electrical response depends on the temperature coefficient of resistance of the active element; this quantity is defined by

$$\alpha \equiv \frac{1}{R}\frac{dR}{dT}. \tag{13.19}$$

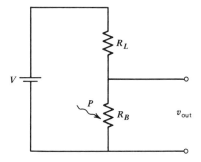

Figure 13.2. Electrical circuit illustrating the operation of a bolometer. R_B and R_L denote the resistances of the bolometer and the load resistor, respectively.

Thus, if the incident radiation causes the temperature of the element to increase by an amount ΔT, the electrical resistance increases by an amount

$$\Delta R = \alpha R_B \Delta T. \tag{13.20}$$

This leads to an output voltage of magnitude

$$v = i\,\Delta R$$

$$= i\alpha R_B \Delta T, \tag{13.21}$$

where i denotes the bias current passing through the bolometer and is given by

$$i = \frac{V}{(R_L + R_B)}. \tag{13.22}$$

Bolometers may employ either metals or semiconductors as their resistive element. The original bolometer of S. P. Langley [*Nature*, **25**, 14 (1881)] used a platinum ribbon as the resistive element; nickel is also commonly employed. Since near room temperature the resistance of most metals depends linearly on temperature, it is approximately true that

$$\alpha = \frac{1}{T}, \tag{13.23}$$

where T denotes the absolute temperature. At room temperature, α is thus approximately equal to $0.003/°C$ for any metal. Platinum and nickel are selected for use in bolometers because they are very strong and can be used in thin ribbons in order to minimize the heat capacity of the detector.

Semiconductor bolometers often have a larger temperature coefficient of resistance α than do metal bolometers. This is because the number of free carriers in a semiconductor is roughly proportional to the Boltzmann factor, which is strongly temperature dependent. The resistance of semiconductor bolometers then changes with temperature as

$$R = R_0 e^{A/T}, \tag{13.24}$$

and α is thus given by

$$\alpha = -\frac{A}{T^2}. \tag{13.25}$$

In a typical semiconductor, A might equal 15,000 K, giving a value of α at room temperature of 0.056. The increased responsivity of semiconductor bolometers does not in any fundamental way improve the detection sensitivity, because in principle the NEP of any thermal detector is limited by temperature noise to the value given by Eq. (13.14). Increased responsivity makes this theoretical limit easier to achieve, however, by lessening the requirements on the electronics used to measure the output voltage.

Semiconductors are characterized by a negative value of α; that is, their resistance decreases as this temperature increases. Semiconductor bolometers are thus subject to a problem known as burnout: If operated with a sufficiently large bias voltage, Joule heating can lower the bolometer's resistance enough that excessive power is dissipated in the device, causing it to be destroyed. This occurrence can be prevented by operating the device at a constant bias current, for example, by placing a large ballast resistor in series with the device.

Increased sensitivity can often be obtained by operating bolometers at low temperatures. The most commonly used low-temperature bolometers are polycrystalline carbon in the form of flakes and single-crystal germanium.

If detectors with a fast temporal response are needed, the resistive element can be placed in direct contact (perhaps by evaporation) with a substate. Such detectors are sometimes called solid backed. This direct contact leads to a large thermal conductance $\mathcal{G}$ and, by Eq. (13.6), to a short response time. However, a large value of $\mathcal{G}$ leads, by Eq. (13.14), to a large value of the NEP, indicating poor device sensitivity.

A small value of the NEP, indicating good sensitivity to weak signals, can be obtained at the expense of good temporal response by supporting the active element in vacuum by only its electrical leads. This provides thermal conductance only by radiation and leads to a value of $\mathcal{G}$ given by Eq. (13.17). Such a device is called vacuum backed.

The NEP of a bolometer may be worse than that of an ideal thermal detector because of Johnson noise introduced by the detector itself or by other components of the detection circuit. The total noise voltage appearing at the output of the circuit in Fig. 13.2 is thus the sum of the contributions due to Johnson noise and to temperature noise, and it is given by

$$\overline{v_N^2} = 4kTR\,\Delta f + i^2\alpha^2 R_B^2 \overline{(\Delta T)^2}(f)\,\Delta f, \qquad (13.26)$$

or, using Eq. (13.2) and assuming that $2\pi f\tau_T \ll 1$, by

$$\overline{v_N^2} = 4kTR\,\Delta f + i^2\alpha^2 R_B^2 \frac{4kT^2}{\mathcal{G}}\,\Delta f. \qquad (13.27)$$

In these expressions, R denotes the parallel combination of R_B and R_L, and it has been assumed that the load resistor and the detector are at the same temperature T. The detector can be considered ideal if the Johnson noise contribution is negligible in comparison to the temperature noise contribution; this condition requires that

$$\frac{R\mathcal{G}}{i^2\alpha^2 R_B^2 T_0} \ll 1.$$

If the condition is fulfilled, the NEP of the detector is given by Eq. (13.18a). In the opposite limit, in which the Johnson noise contribution is much greater than that of temperature noise, the NEP is determined by noting that the signal voltage produced by constant power P_0 falling onto the detector is given using Eqs. (13.9) and (13.21) as

$$v_s = \frac{\varepsilon i\alpha R_B P_0}{\mathcal{G}}. \qquad (13.28)$$

The value of P_0 for which the square of v_S is equal to v_n^2 gives the NEP as

$$P_N = \frac{\sqrt{4kT_0 R\,\Delta f}}{\varepsilon i\alpha R_B/\mathcal{G}}. \qquad (13.29)$$

13.3 PYROELECTRIC DETECTORS

Pyroelectric detectors have been developed fairly recently. They are reasonably sensitive, are usually operated at room temperature, and display an unusually good frequency response for a thermal detector, extending to as much as several hundred megahertz.

Pyroelectric detectors are made from *ferroelectric* crystals, that is, crystals that can exhibit a permanent electric dipole moment even in the absence of an applied electric field. [Ferroelectricity is discussed in C. Kittel, *Introduction to Solid State Physics*, Wiley, New York, 1976, Chapter 13.] Commercially available pyroelectric detectors are made of triglycene sulfate, lithium tantalate, strontium barium niobate, and polyvinylidene fluoride. At room temperature, the magnitude of the polarization is often found to be strongly temperature dependent. Thus, as the temperature of the detector is varied, the electric dipole moment of the crystal must change, leading to the motion of bound charge. If electrodes are placed on the surfaces of the crystal, the motion of bound charge within the crystal can induce the flow of current through an external circuit. The magnitude of this current is given by

$$i = pA\frac{dT}{dt},\qquad(13.30)$$

where A is the area of the electrode and where p is the pyroelectric coefficient, which might typically have the value 3×10^{-8} C/cm^2 K. Thus a pyroelectric detector does not respond to constant input power, since only the time derivative of the detector's temperature leads to a signal current. Let us assume then that the power falling onto the detector has been modulated, and that, as before, one harmonic component of the power is represented by the real part of

$$P(t) = P_0 e^{i2\pi ft}.$$

Using Eq. (13.8) for the temperature rise T_1 of the detector element, the time derivative dT/dt is given by

$$\frac{dT}{dt} = \frac{i2\pi f\varepsilon P_0}{\mathcal{G} + i2\pi f\mathcal{C}}e^{i2\pi ft}.\qquad(13.31)$$

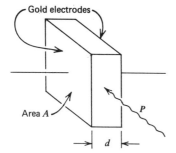

Gold electrodes

Area A

P

d

Figure 13.3. Pyroelectric detector.

Thus, ignoring the possible phase shift between $T_1(t)$ and $P(t)$, the current responsivity is given by

$$|\mathscr{R}_i| = \frac{2\pi f p A \varepsilon / \mathscr{G}}{\left[1 + (2\pi f \tau_T)^2\right]^{1/2}} . \tag{13.32}$$

A typical circuit for use with a pyroelectric detector is shown in Fig. 13.4. Here i represents an ideal current source whose strength is given by Eq. (13.30) and R_i and C_i represent the internal shunt resistance and capacitance of the detector, respectively. The current i produces an output voltage whose magnitude depends on the values of the load resistance R_L and any capacitance C_L of the external circuit. This voltage depends on the resistance

$$R = \frac{R_i R_L}{R_i + R_L}$$

and capacitance

$$C = C_i + C_L$$

of the parallel combination of components. This circuit is thus characterized by an electrical response time

$$\tau_E = RC,$$

leading to a voltage responsivity of the form

$$|\mathscr{R}_v| = |\mathscr{R}_i| \frac{R}{\left[1 + (2\pi f \tau_E)^2\right]^{1/2}} , \tag{13.33}$$

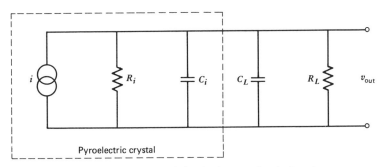

Figure 13.4. Equivalent electrical circuit of a pyroelectric detection system.

or, using Eq. (13.32), as

$$|\mathcal{R}_v| = \frac{2\pi f p A \varepsilon R / \mathcal{G}}{\left[1 + (2\pi f \tau_T)^2\right]^{1/2}\left[1 + (2\pi f \tau_E)^2\right]^{1/2}}. \tag{13.34}$$

The frequency dependence of the responsivity predicted by Eq. (13.34) is shown graphically in Fig. 13.5. These curves assume values of the parameters appearing in the equation that are typical of commercially available pyroelectric detectors: $\tau_T = 0.016$ s, implying a thermal cutoff frequency $f_{cT} \equiv 1/2\pi\tau_T = 10$ Hz; $\varepsilon p A / G \tau_T = 1$ μA/W; device capacitance $C = 16$ pF. Several different values of the resistance are illustrated, leading to different values of $\tau_E = RC$ and of the electrical cutoff frequency $f_{cE} \equiv 1/2\pi RC$. For modulation frequencies in the range $f_{cT} \ll f \ll f_{cE}$, the current responsivity is essentially flat. The voltage responsivity increases with the value of the load resistance R, but the electrical cutoff frequency is also reduced by the increase in the RC time constant.

Johnson noise is the predominant noise source in most pyroelectric detectors. The rms noise voltage is then

$$v_N = \sqrt{4kTR\,\Delta f}, \tag{13.35}$$

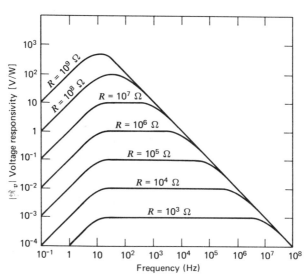

Figure 13.5. Frequency response of a typical pyroelectric detector for several values of the load resistance.

and the NEP is thus given by

$$P_N = \frac{v_N}{|\mathcal{R}_v|}. \tag{13.36}$$

In the frequency range $f_{cT} \ll f \ll f_{cE}$, the voltage responsivity is given by

$$|\mathcal{R}_v| = \frac{pA\varepsilon R}{\mathcal{G}\tau_T}, \tag{13.37}$$

and thus for these frequencies, the NEP is given by

$$P_N = \frac{\sqrt{4kT\Delta f/R}}{pA\varepsilon/\mathcal{G}\tau_T}. \tag{13.38}$$

The NEP can thus be improved by increasing the value of R, but only at the expense of a decreased frequency response. For the numerical example given earlier, the largest value of R consistent with $f_{cT} \ll f_{cE}$ is approximately $R = 10^8 \ \Omega$. Equation (13.38) then implies an NEP referred to a 1-Hz bandwidth of 10^{-8} W, which is typical for commercial pyroelectric detectors.

BIBLIOGRAPHY

N. Coron, *Infrared Phys.* **16**, 411 (1976).

R. C. Jones, in *Advance in Electronics*, Volume 5, Academic, New York, 1953.

R. H. Kingston, *Detection of Optical and Infrared Radiation*, Springer, Berlin, 1978, Chapter 7.

P. W. Kruse, L. D. McGlauchlin, and R. B. McQuistan, *Infrared Technology*, Wiley, New York, 1962.

E. H. Putley, in *Optical and Infrared Detectors*, R. J. Keyes, ed., Springer, Berlin, 1977.

R. A. Smith, F. E. Jones, and R. P. Chasmar, *The Detection and Measurement of Infrared Radiation*, Oxford University, London, 1968.

14

Fluctuations of the Radiation Field

The power falling onto a detector exposed to thermal radiation (by thermal radiation we mean radiation having the same statistical properties as spectrally filtered blackbody radiation) is not constant, but fluctuates due to the stochastic nature of the radiation field. These fluctuations are impressed on the electrical output of the detector and can be considered an additional source of noise in the detection process. For the case of a photon detector, these fluctuations give rise to noise in the photocurrent in excess of *shot noise*, which is due to the discreteness of the radiation field (i.e., to the fact that the field is comprised of photons) and which is present even in the detection of radiation of constant power. This *excess noise* in the photocurrent is significant only if the photon occupation number of the radiation field is appreciably greater than unity. For the case of a thermal detector, the power fluctuations in the radiation act along with the discreteness of absorbed and emitted radiation to produce *temperature noise*, that is, fluctuations in the temperature of the active element of the detector.

The present chapter first presents the classical theory of fluctuations of thermal radiation and then, in Section 14.2, shows how these fluctuations can become impressed on the electrical response of a photon detector. Section 14.3 presents a discussion of power fluctuations within the context of a quantized radiation field. In this context, noise is introduced into the detection process by the randomness in the arrival times of individual photons. This noise is known as *photon noise* and includes both shot-noise and excess-noise contributions. Power fluctuations, which give rise to the excess noise, are in this framework a consequence of the Bose–Einstein nature of photons, which causes the arrival times of individual photons to show a positive correlation (i.e., to be bunched). As an illustration of the effect of power fluctuations on detection sensitivity, Section 14.4 discusses the photon-noise-limited noise equivalent power (NEP) of ideal thermal and photon detectors.

14.1 STATISTICS OF A CLASSICAL THERMAL SOURCE

Let us consider the radiation field produced at a detector by a distant thermal source. The analysis of the statistical properties of this radiation is simplified if we assume that the radiation is *quasi-monochromatic*, that is, that the radiation is substantially contained in a small spectral interval $\Delta\nu$ centered at some frequency ν_0. It is immaterial whether $\Delta\nu$ is the intrinsic frequency breadth of the source or whether $\Delta\nu$ is limited by a filter placed in front of the detector. We shall also assume for the present that the source of the radiation is essentially a point source, so that the radiation is spatially coherent over the active area of the detector. This will be the case if the angle subtended by the source is much less than λ/D, where λ is the characteristic wavelength of the field being considered and where D is a characteristic dimension of the detector. Thus, in the language introduced in Chapter 12, the detector receives radiation from a single transverse mode of the radiation field.

The electric field $E(t)$ measured at some point on the detector can be described by the real part of the quantity

$$\hat{E}(t)e^{i2\pi\nu_0 t}. \tag{14.1}$$

Purely monochromatic radiation is characterized by a constant value of the complex amplitude $\hat{E}(t)$, but for the case of thermal radiation $\hat{E}(t)$ is a fluctuating quantity. These fluctuations arise because a thermal source contains a large number of independent radiators, and the field amplitude $\hat{E}(t)$ is an incoherent superposition of the contributions of each of these radiators. Temporal changes in the amplitudes and relative phases of these radiators, which may be due to collisions within the thermal source, thus lead to fluctuations in $\hat{E}(t)$.

The stochastic nature of the field amplitude can be visualized by considering $\hat{E}(t)$ to be a vector in the complex E plane, as shown in Fig. 14.1. The real and imaginary parts of $\hat{E}(t)$ fluctuate randomly and independently of one another. Since each part is the sum of a large number of independent contributions, each part must, as a consequence of the central limit theorem (see, e.g., p. 81 of the book by Davenport and Root, listed in the bibliography at the end of this chapter), have a Gaussian probability distribution. The joint probability distribution of the complex random variable $\hat{E}(t)$ thus has the form of a two-dimensional Gaussian, as shown in Fig. 14.2. The analytic form of this distribution is most conveniently expressed in terms of the length E and phase ϕ of $\hat{E}(t)$, defined in Fig. 14.1, and is given by

$$p(E, \phi) = \frac{1}{\pi E_0^2} e^{-E^2/E_0^2}. \tag{14.2}$$

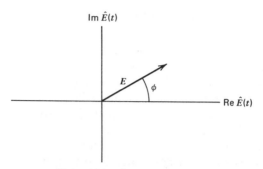

Figure 14.1. Complex E plane.

This probability distribution has the interpretation that the probability that the tip of the vector $\hat{E}(t)$ lies within the area element $E\,dE\,d\phi$ surrounding the point with polar coordinates (E, ϕ) is given by $p(E, \phi)E\,dE\,d\phi$. The distribution is normalized in the sense that

$$\int_0^{2\pi} d\phi \int_0^{\infty} dE\, E\, p(E, \phi) = 1. \tag{14.3}$$

The parameter E_0 has been introduced here as a phenomenological measure of the variance of the probability distribution. Its physical significance can be obtained by calculating the expectation value of E^2:

$$\langle E^2 \rangle = \int_0^{2\pi} d\phi \int_0^{\infty} dE\, E^3 p(E, \phi)$$

$$= 2E_0^2 \int_0^{\infty} \frac{dE\, E^3}{E_0^4} e^{-E^2/E_0^2}$$

$$= E_0^2. \tag{14.4}$$

We can define the physical intensity* of the radiation as the magnitude of the Poynting vector averaged over one cycle of the field; using Eq. (1.25), we obtain for the intensity

$$I = \tfrac{1}{2} c\varepsilon_0 E^2. \tag{14.5}$$

---

*In the present context, the quantity given by Eq. (14.5) is usually referred to as the intensity (or physical intensity) of the radiation. This quantity bears no obvious relation to the quantity introduced in Eq. (2.7), which is also known as the intensity (or as the radiometric intensity).

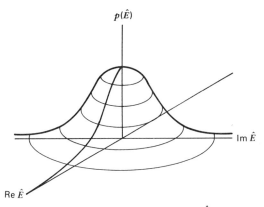

Figure 14.2. Probability distribution of the complex amplitude $\hat{E}$ for a single-transverse-mode thermal radiation field.

The average value of the intensity is thus proportional to E_0^2 and is given by

$$\langle I \rangle \equiv I_0 = \tfrac{1}{2} c \varepsilon_0 E_0^2. \tag{14.6}$$

In analyzing the detection process, we are usually more interested in the statistical properties of the intensity than in those of the field. The probability distribution for I can be obtained by first noting that the probability that the value of I is less than some fixed value I' is given by

$$P(I < I') = \int_0^{2\pi} d\phi \int_0^{\sqrt{2I'/\varepsilon c}} dE \, E p(E, \phi) = (1 - e^{-I'/I_0}). \tag{14.7}$$

The probability density of I' is thus given by

$$p(I') = \frac{d}{dI'} P(I < I')$$

$$= \frac{1}{I_0} e^{-I'/I_0},$$

or, formally substituting I for I', by

$$p(I) = \frac{1}{I_0} e^{-I/I_0}. \tag{14.8}$$

This probability distribution is shown in Fig. 14.3. The most probable value

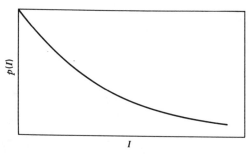

Figure 14.3. Probability distribution of the intensity of a single-transverse-mode thermal radiation field.

of the intensity is thus zero. This distribution is normalized since

$$\int_0^\infty p(I)\, dI = 1.$$

We shall often be concerned with the moments of this distribution, given by

$$\langle I^n \rangle \equiv \int_0^\infty I^n p(I)\, dI$$

$$= I_0^n \int \left(\frac{I}{I_0} \right)^n e^{-I/I_0} \frac{dI}{I_0}$$

$$= n! I_0^n. \tag{14.9}$$

For instance, since

$$\langle I \rangle = I_0$$

and

$$\langle I^2 \rangle = 2 I_0^2, \tag{14.10}$$

the variance of I is given by

$$\langle (\Delta I)^2 \rangle = \langle (I - \langle I \rangle)^2 \rangle$$

$$= I_0^2. \tag{14.11}$$

Therefore, the rms fluctuation of I is given by

$$(\Delta I)_{\text{rms}} = I_0. \tag{14.12}$$

This result tells us that classical thermal radiation is extremely noisy, being characterized by 100% intensity fluctuations. [Recall that in the limit $\bar{n} \gg 1$, the Planck distribution implies that the dispersion in photon occupancy is given by $(\Delta n)_{rms} = \bar{n}$ (cf. Eq. 3.64), leading to intensity fluctuations given by Eq. (14.12). This point of view will be further exploited in the following section.]

Spectral Density of Fluctuations

We have not yet mentioned the time scale over which these fluctuations occur. If the radiation field is largely composed of frequency components contained within a spectral interval $\Delta\nu$ centered on the frequency ν_0, we expect the field to display temporal coherence over times of the order of the coherence time taken as

$$\tau_c \approx \frac{1}{\Delta\nu},$$

and we would expect no fluctuations to occur on a time scale much shorter than τ_c. Figure 14.4 shows, for instance, a computer simulation of $I(t)$. This

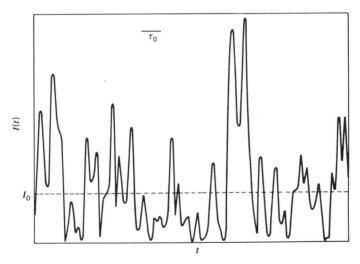

Figure 14.4. Time dependence of the intensity of a single-transverse-mode thermal radiation field, obtained from a computer simulation of collision broadening of an atomic emission line. τ_0 denotes the mean time between collisions. Adapted, with permission, from R. Loudon, *The Quantum Theory of Light*, Clarendon Press, Oxford, 1973.

curve displays the expected behavior with respect to the coherence time, and it also displays the large fractional fluctuations characteristic of thermal radiation.

The exact relationship between intensity, or power, fluctuations and the spectral composition of thermal radiation can be formulated by means of the Wiener–Khintchine relations discussed in Chapter 8. We let $P(\nu)$ designate the spectral density of the radiation field, that is, the power per unit frequency interval incident on the detector from a single transverse field mode. The normalized spectral density, also known as the line-shape function, can then be defined by

$$g(\nu) = \frac{P(\nu)}{\int_0^\infty P(\nu)\, d\nu}, \tag{14.13}$$

and clearly obeys the relation

$$\int_0^\infty g(\nu)\, d\nu = 1.$$

By the Wiener–Khintchine relations, used here in the form of Eqs. (8.17) and (8.18), the (first-order) field correlation function defined by

$$\gamma(\tau) = \frac{\langle E(t)E(t+\tau)\rangle}{\langle E(t)^2\rangle} \tag{14.14}$$

is related to $g(\nu)$ by

$$\gamma(\tau) = \int_0^\infty g(\nu)\cos 2\pi\nu\tau\, d\nu \tag{14.15}$$

and by the inverse relation

$$g(\nu) = 4\int_0^\infty \gamma(\tau)\cos 2\pi\nu\tau\, d\tau. \tag{14.16}$$

The field correlation function $\gamma(\tau)$ displays oscillations at the field central frequency and can always be represented as

$$\gamma(\tau) = g^{(1)}(\tau)\cos 2\pi\nu\tau. \tag{14.17}$$

This relation defines the first-order coherence function $g^{(1)}(\tau)$.

Formal definitions of the coherence time τ_c and the equivalent spectral breadth $\Delta\nu$ of the thermal field can be given in terms of the quantities $g(\nu)$

and $g^{(1)}(\tau)$ just introduced. If τ_c is defined by

$$\tau_c = \int_{-\infty}^{\infty} \left[g^{(1)}(\tau) \right]^2 d\tau \tag{14.18}$$

and if $\Delta \nu$ is defined by

$$(\Delta \nu)^{-1} = \int_0^{\infty} \left[g(\nu) \right]^2 d\nu, \tag{14.19}$$

then, by Parseval's theorem, these two quantities are related by

$$\Delta \nu \, \tau_c = 1.$$

[These definitions of τ_c and $\Delta \nu$ are due to L. Mandel, *Proc. Phys. Soc. London*, **72**, 1037 (1958).]

The fluctuations in optical power can similarly be described by a normalized power correlation function (or second-order field correlation function) defined by

$$g^{(2)}(\tau) = \frac{\langle P(t)P(t+\tau) \rangle}{\langle P(t) \rangle^2}. \tag{14.20}$$

so long as the fluctuations in $E(t)$ obey a Gaussian probability distribution, as they do for thermal radiation, $g^{(2)}(\tau)$ is related by $g^{(1)}(\tau)$ by

$$g^{(2)}(\tau) = 1 + \left[g^{(1)}(\tau) \right]^2. \tag{14.21}$$

[See, e.g., Eq. (5.101) of R. Loudon, *The Quantum Theory of Light*, Oxford University, London, 1973]. This result can also be considered a special case of the Gaussian moment theorem, discussed on p. 18 of the book by Saleh listed in the bibliography at the end of this chapter. The power spectrum $S_p(\nu)$ of the intensity fluctuations can be obtained through an additional application of the Wiener–Khintchine relation (8.17) to give

$$S_P(\nu) = 4\langle P(t) \rangle^2 \int_0^{\infty} g^{(2)}(\tau) \cos 2\pi \nu \tau \, d\tau. \tag{14.22}$$

By inserting Eq. (14.21) into the integral and making use of the result $\langle P^2 \rangle = 2\langle P \rangle^2$ (see Eq. 14.10), this result becomes

$$S_P(\nu) = \langle P(t)^2 \rangle \delta(\nu) + \langle P(t)^2 \rangle \int_{-\infty}^{\infty} \left[g^{(1)}(\tau) \right]^2 \cos 2\pi \nu \tau \, d\tau. \tag{14.23}$$

There is thus a dc contribution, corresponding to the average value of the power, and a contribution that explicitly depends on the spectral distribution of the radiation.

Examples. These results can perhaps best be illustrated using specific cases. Consider first the case of a radiation field whose normalized spectral density has the form of the Lorentzian line shape

$$g(\nu) = \frac{1}{\pi} \left[\frac{1/2\pi\tau_0}{(\nu - \nu_0)^2 + (1/2\pi\tau_0)^2} \right]. \tag{14.24}$$

Such a line shape occurs, for instance, for the case of a gas of collisionally broadened atoms, if ν_0 is the transition frequency and τ_0 is the mean time between collisions. The normalized field correlation function is then given by Eq. (14.15) as

$$\gamma(\tau) = e^{-|\tau|/\tau_0} \cos 2\pi\nu_0\tau, \tag{14.25}$$

and the first-order coherence function is given by Eq. (14.17) as

$$g^{(1)}(\tau) = e^{-|\tau|/\tau_0}. \tag{14.26}$$

These results are illustrated in Fig. 14.5a. Correlations are seen to die out in a time comparable to τ_0. By using the form (14.25) for $\gamma(\tau)$ in Eq. (14.18), it can be seen that the coherence time τ_c is equal to the parameter τ_0. Using the general result (14.21), the second-order degree of coherence is given by

$$g^{(2)}(\tau) = 1 + e^{-2|\tau|/\tau_0}, \tag{14.27}$$

which is illustrated in Fig. 14.5b. The correlation is again seen to decay in a time comparable to τ_0, but now reaches a minimum value of unity. The spectral density of the power fluctuations is given by evaluating Eq. (14.23) for this form of $g^{(2)}(\tau)$, giving

$$S_P(\nu) = \langle P(t)^2 \rangle \delta(\nu) + \frac{2}{\pi} \langle P(t)^2 \rangle \left(\frac{1/\pi\tau_0}{\nu^2 + (1/2\pi\tau_0)^2} \right). \tag{14.28}$$

This relation is illustrated in Fig. 14.6.

As a second example, consider a nonfluctuating optical signal, such as the classical stable wave given by

$$E(t) = E_0 \cos 2\pi ft. \tag{14.29}$$

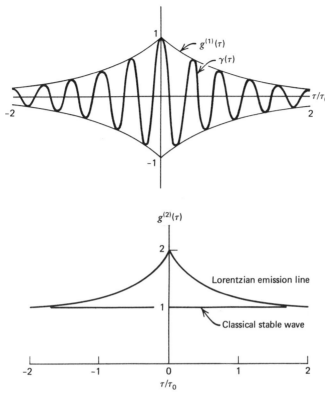

Figure 14.5. (*a*) Normalized field correlation function $\gamma(\tau)$ and first-order field coherence function $g^{(1)}(\tau)$ for a Lorentzian emission line. For clarity, the period of the oscillations in $g^{(1)}(\tau)$ is greatly exaggerated. (*b*) Second-order field coherence function $g^{(2)}(\tau)$ for a Lorentzian emission line. For comparison, $g^{(2)}(\tau)$ for a classical stable wave is also shown.

For such a signal, the normalized autocorrelation function is given by

$$\gamma(\tau) = \cos 2\pi f t, \tag{14.30}$$

and the first-order coherence function is given by

$$g^{(1)}(\tau) = 1 \tag{14.31}$$

for all values of τ. The classical stable wave does not show any fluctuations, and thus directly from the definition (14.20) of $g^{(2)}(\tau)$ the second-order coherence function is given by

$$g^{(2)}(\tau) = 1.$$

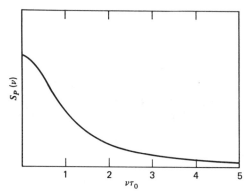

Figure 14.6.   Spectral density of the power fluctuations of a Lorentzian emission line.

14.2   EFFECT OF POWER FLUCTUATIONS ON NOISE IN PHOTON DETECTION

Let us now examine the effect of power fluctuations, whose correlation function and spectral density are given by Eqs. (14.21) and (14.22), on the photon detection process. In particular, we shall calculate the noise introduced into the photocurrent by power fluctuations. This discussion will be a generalization of the derivation of photon noise given in Section 8.2 in that we are relaxing the assumption made there that constant power falls onto the detector. We assume, as in Section 14.1, that the radiation is quasi-monochromatic and spatially coherent across the detector (i.e., that it is obtained from a single transverse field mode).

We assume that the probability that a photoelectron is ejected in the time interval t to $t + dt$ is given by

$$p_1(t, dt) = \alpha P(t)\, dt. \tag{14.32}$$

Here $P(t)$ is the instantaneous value of the optical power (averaged over one cycle of the field oscillation) falling onto the detector, and the notation $\alpha = \eta/h\nu$ has been introduced. The average number of photoelectrons emitted in a finite time interval running from t to $t + T$ is thus given by

$$\mu = \alpha \int_t^{t+T} P(t)\, dt \tag{14.33a}$$

or equivalently as

$$\mu = \alpha \bar{P}(t) T, \tag{14.33b}$$

where

$$\bar{P}(t) = \frac{1}{T} \int_t^{t+T} P(t)\, dt. \tag{14.34}$$

In the ensuing discussion, we shall consistently use the bar rotation to refer to averages performed over a time interval T, and we shall use angular bracket notation to refer to long-time averages (or to ensemble averages due to ergodicity), for instance,

$$\langle P(t) \rangle = \lim_{T' \to \infty} \frac{1}{T'} \int_{-T'/2}^{T'/2} P(t)\, dt.$$

The probability that N photoelectrons are emitted in the time interval T is given by the Poisson distribution [Eq. (8.32)]

$$p_N(t, T) = \frac{\mu^N}{N!} e^{-\mu}.$$

This probability distribution is itself a random function of the time t, however, since the power $\bar{P}(t)$ and hence μ are fluctuating quantities. In calculating the mean and variance of the number of photoevents occurring in an interval of length T, averaged over all possible values of the initial time t, these fluctuations must be taken into account. The average of the mean number of photoevents in the time interval T is thus given by

$$\langle \bar{N} \rangle = \langle \mu \rangle = \alpha \langle \bar{P}(t) \rangle T, \tag{14.35}$$

where the sharp brackets denote an average over all possible values of the initial time t. From the properties of the Poisson distribution [see, e.g., Eqs. (8.36) and (8.37)] the expectation value of the square of N is given by

$$\overline{N^2} = \mu + \mu^2, \tag{14.36}$$

and the long-time average of this quantity is thus given by

$$\langle \overline{N^2} \rangle = \langle \mu \rangle + \langle \mu^2 \rangle. \tag{14.37}$$

The long-time average of the variance of the number of photoevents in the time interval T is then given by

$$\langle \overline{(\Delta N)^2} \rangle = \langle \overline{N^2 - \bar{N}^2} \rangle$$

$$= \langle \mu \rangle + \langle \mu^2 \rangle - \langle \mu \rangle^2. \tag{14.38}$$

In the absence of power fluctuations, $\langle \mu^2 \rangle$ equals $\langle \mu \rangle^2$, and the long-time-averaged variance is given by

$$\overline{\langle (\Delta N)^2 \rangle} = \langle \mu \rangle = \langle \overline{N} \rangle, \tag{14.39}$$

which is in agreement with Eq. (8.38). In the presence of power fluctuations, this result takes the more general form

$$\overline{\langle (\Delta N)^2 \rangle} = \alpha \langle \overline{P}(t) \rangle T$$

$$+ \alpha^2 T^2 \langle (\Delta \overline{P})^2 \rangle, \tag{14.40}$$

where $\langle (\Delta \overline{P})^2 \rangle$ denotes the variance of the power fluctuations averaged over an integration time T, that is,

$$\langle (\Delta \overline{P})^2 \rangle = \langle \overline{P}^2 \rangle - \langle \overline{P} \rangle^2. \tag{14.41}$$

The long-time average of $\overline{P}^2$ can be calculated as

$$\langle \overline{P}^2 \rangle = \left\langle \left[\frac{1}{T} \int_T^{t+T} P(t') \, dt' \right]^2 \right\rangle$$

$$= \left\langle \frac{1}{T^2} \int_t^{t+T} dt' \int_t^{t+T} dt'' \, P(t') P(t'') \right\rangle$$

$$= \left\langle \frac{1}{T^2} \int_t^{t+T} dt' \int_t^{t-t'+T} d\tau \, P(t') P(t'+\tau) \right\rangle, \tag{14.42}$$

where the substitution $t'' = t' + \tau$ was used in obtaining the last line. This result can be reexpressed as

$$\langle \overline{P}^2 \rangle = \frac{1}{T^2} \int_0^T dt' \int_{-t'}^{-t'+T} d\tau \, \langle P(t') P(t'+\tau) \rangle$$

or, using Eqs. (14.20) and (14.21), as

$$\langle \overline{P}^2 \rangle = \frac{\langle \overline{P} \rangle^2}{T^2} \int_0^T dt' \int_{-t'}^{-t'+T} d\tau \left\{ 1 + [g^{(1)}(\tau)]^2 \right\}$$

$$= \frac{\langle \overline{P} \rangle^2}{T^2} \int_0^T dt' \left\{ T + \int_{-t'}^{-t'+T} d\tau \, [g^{(1)}(\tau)]^2 \right\}. \tag{14.43}$$

In general, the remaining integration over the variable τ cannot be performed in closed form. In detection systems, the integration time T is usually much greater than the coherence time τ_c defined by Eq. (14.18). Since the coherence time can be interpreted as the time interval over which $(g^{(1)}(\tau))^2$ has an appreciable magnitude, in the limit $\tau_c \ll T$ the limits of integration in the τ integration can be replaced by $\pm \infty$, allowing $\langle \bar{P}^2 \rangle$ to be expressed as

$$\langle \bar{P}^2 \rangle = \frac{\langle \bar{P} \rangle^2}{T^2} \int_0^T dt' \, (T + \tau_c)$$

$$= \langle \bar{P} \rangle^2 \left(1 + \frac{\tau_c}{T} \right) \tag{14.44}$$

Finally, using Eq. (14.41), the mean-square power fluctuations are given by

$$\langle (\Delta \bar{P})^2 \rangle = \langle \bar{P} \rangle^2 \frac{\tau_c}{T} \quad \text{for } \tau_c \ll T. \tag{14.45}$$

This result was first obtained by Purcell [E. M. Purcell, *Nature*, **178**, 1449 (1956)]. This result reflects the fact that power fluctuations tend to average to zero for integrations times much longer than τ_c. In particular, the variance of $\bar{P}$ decreases linearly with the number (T/τ_c) of statistically independent measurements performed on $P(t)$ during an integration of length T. Using this result, the variance of N is obtained from Eq. (14.40) as

$$\langle \overline{(\Delta N)^2} \rangle = \alpha T \langle P(t) \rangle + \alpha^2 T^2 \langle \bar{P} \rangle^2 \frac{\tau_c}{T}$$

$$= \langle \bar{N} \rangle + \langle \bar{N} \rangle^2 \frac{\tau_c}{T}$$

$$= \langle \bar{N} \rangle \left(1 + \langle \bar{N} \rangle \frac{\tau_c}{T} \right). \tag{14.46}$$

Since the photocurrent is related in the number of photoevents by

$$i(t) = \frac{eN}{T}, \tag{14.47}$$

it is possible to relate the stochastic properties of N to those of the photocurrent. The time-averaged photocurrent is thus given by

$$\bar{i} = \frac{e \langle \bar{N} \rangle}{T} = \frac{\eta e}{h\nu} \langle \bar{P}(t) \rangle, \tag{14.48}$$

and the mean-square current noise is given by

$$\overline{i_N^2} = \frac{e^2 \langle \overline{(\Delta N)^2} \rangle}{T^2}$$

$$= \frac{e\bar{i}}{T} + \bar{i}^2 \frac{\tau_c}{T}, \tag{14.49}$$

or, introducing $\Delta f = 1/2T$ (cf. Eq. 7.18), by

$$\overline{i_N^2} = 2e\bar{i}\,\Delta f + \bar{i}^2 2\tau_c\,\Delta f$$

$$= 2e\bar{i}\,\Delta f \left(1 + \frac{\langle \overline{N} \rangle \tau_c}{T}\right). \tag{14.50}$$

This result shows that the total noise is greater than shot noise (cf. Eq. 8.42) by a factor of one plus the average number of photoevents per coherence time.

14.3 PHOTON NOISE

Sections 14.1 and 14.2 described the fluctuations of a thermal radiation field from a purely classical point of view. The discussion included quantum mechanical effects only insofar as the ejection of a photoelectron was assumed to be a stochastic process whose rate was given by Eq. (14.32).

The present section discusses field fluctuations from a purely quantum mechanical point of view. The analysis given here is in many ways less detailed than that given in the preceding sections, but the mathematics is also considerably simpler than that required by the classical description. An advantage of the quantum mechanical description is that shot noise and excess noise are treated in a unified fashion. Thus they are seen to be manifestations of the same effect, that is, the randomness of the arrival times of the photons at the detector. The present analysis is therefore applicable to either thermal or photon detectors, and it tends to minimize the somewhat artificial distinction between the inherent noise limitations of these two types of detectors. In addition, the present treatment is capable of treating spatially incoherent (i.e., multi-transverse-mode) fields of large frequency breadth.

The mean number of photons contained in a mode of a thermal radiation field was shown in Eq. (3.55) to be

$$\bar{n} = \frac{1}{e^{h\nu/kT} - 1}, \tag{14.51}$$

where ν is the frequency of the radiation and where T is a temperature that characterizes the excitation mechanism. As an example, for blackbody radiation, T is the physical temperature of the walls of the radiation enclosure. The actual number of photons associated with a field mode fluctuates from this mean value; the variance in this number is given by [see Eq. (3.62)]

$$\overline{(\Delta n)^2} \equiv (\Delta n)^2_{\mathrm{rms}} \equiv \overline{(n - \bar{n})}^2$$

$$= \bar{n}(\bar{n} + 1), \tag{14.52}$$

which can be expressed using Eq. (14.51) as

$$\overline{(\Delta n)^2} = \frac{e^{h\nu/kT}}{\left(e^{h\nu/kT} - 1\right)^2}. \tag{14.53}$$

It is possible to characterize noise in the photon detection process as resulting from the fluctuations in the rate at which photons strike the detector. For the case of a unit-quantum-efficiency detector, all the incident photons produce a detection event. Thus the noise in the photocurrent can be considered to result solely from noise in the radiation field. A detector with a quantum efficiency that is less than unity will introduce additional noise into the photocurrrent, since stochastic processes determine whether or not a particular incident photon produces a photoevent. This increased noise is called partition noise. It is possible to model an ideal photon detector of quantum efficiency $\eta < 1$ as an ideal, unit-quantum-efficiency detector placed behind a filter of transmittance η. The mean number of photons reaching the detector is then $\eta\bar{n}$, and the variance in this number is then $\eta\bar{n}(\eta\bar{n} + 1)$. [This simple model of partition noise can fail for the case of a photoconductor in radiative equilibrium with the incident radiation, as has been discussed by K. M. van Vliet, *Appl. Opt.* **6**, 1145 (1967)].

The average value of the thermal power falling onto a detector of area A from a projected solid angle Ω_{proj} in a small frequency band $\Delta\nu$ (as illustrated in Fig. 14.7) is given in terms of the spectral radiance L_ν of the source by

$$\bar{P} = L_\nu A \Omega_{\mathrm{proj}} \Delta\nu$$

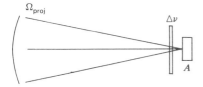

Figure 14.7. Radiation of spectral breadth $\Delta\nu$ falls onto a detector of area A from a projected solid angle Ω_{proj}.

or, using Eq. (3.69) for L_ν, as

$$\bar{P} = \frac{2 A \Omega_{\text{proj}} \Delta\nu}{c^2} \frac{h\nu^3}{e^{h\nu/kT} - 1} . \tag{14.54}$$

Using Eq. (14.51), this result can be expressed as

$$\bar{P} = h\nu\bar{n}r, \tag{14.55}$$

where

$$r = 2\left(\frac{A\Omega_{\text{proj}}}{\lambda^2}\right) \Delta\nu \tag{14.56}$$

can be interpreted as the rate at which field modes intersect the detector, as is discussed more completely in Section 12.3. Since the number of photons per mode is a fluctuating quantity, the instantaneous value of the power also fluctuates. The rms value of the power fluctuations is defined as

$$(\Delta P)_{\text{rms}} = \sqrt{\overline{(P - \bar{P})^2}} \equiv \sqrt{\overline{(\Delta P)^2}} . \tag{14.57}$$

Most detection systems are not sufficiently fast to respond to the instantaneous intensity. Instead, such systems respond to the intensity averaged over an effective integration time $T = (2 \Delta f)^{-1}$ that is much greater than the coherence time $\tau_c = (\Delta\nu)^{-1}$ of the radiation. The rms fluctuation in incident power averaged over a time $T \gg \tau_c$ is given by

$$(\Delta P)_{\text{rms}} = \frac{h\nu(\Delta n)_{\text{rms}} r}{\sqrt{rT}} , \tag{14.58}$$

since $h\nu(\Delta n)_{\text{rms}}$ represents the rms dispersion in the energy per mode, r represents the rate at which these modes fall onto the detector, and the product of these factors has been divided by the square root of the number rT of modes that contribute to the time average, reflecting the reduction in noise through averaging as discussed in Section 7.4. This expression can be expressed in terms of Δf, $\bar{n}$, and $\bar{P}$, giving

$$(\Delta P)_{\text{rms}} = [2h\nu\bar{P}(\bar{n} + 1) \Delta f]^{1/2}. \tag{14.59}$$

In this equation, $\bar{n}$ denotes the photon occupation number of the radiation as measured at the detector. This number can differ from the occupation number measured at the source if the radiation has been attenuated, for instance, by absorption or by diffraction effects.

Power fluctuations of the sort just discussed can give rise to fluctuations in the current produced by a photon detector. We assume that the detector quantum efficiency is equal to η so that the mean value of the photocurrent is given by

$$\bar{i} = \eta \frac{e\bar{P}}{h\nu}. \tag{14.60}$$

Furthermore, the mean-square noise in the photocurrent is given by

$$\overline{i_N^2} = \frac{e^2}{h^2\nu^2}(\Delta P_{\text{abs}})_{\text{rms}}^2. \tag{14.61}$$

Here $(\Delta P_{\text{abs}})_{\text{rms}}$ denotes the rms fluctuation in the absorbed power. As discussed earlier, this quantity can be obtained by replacing $\bar{n}$ with $\eta\bar{n}$ in the expression for $(\Delta n)_{\text{rms}}$ appearing in Eq. (14.58), giving

$$(\Delta P_{\text{abs}})_{\text{rms}} = \sqrt{2h\nu\eta\bar{P}(\eta\bar{n} + 1)\,\Delta f}. \tag{14.62}$$

Equation (14.62) can be used to express Eq. (14.61) as

$$\overline{i_N^2} = 2e\bar{i}\,\Delta f(\eta\bar{n} + 1) \tag{14.63}$$

or, equivalently, as

$$\overline{i_N^2} = 2e\bar{i}\,\Delta f + \frac{\bar{i}^2\,\Delta f}{\left(A\Omega_{\text{proj}}/\lambda^2\right)\Delta\nu}. \tag{14.64}$$

The first term in this expression results from the contribution of unity to the factor $(\eta\bar{n} + 1)$ in Eq. (14.63), and thus this term is dominant in the limit $\eta\bar{n} \ll 1$. This first term corresponds to shot noise in the photocurrent and is present even in the absence of power fluctuations. In the limit $\bar{n} \ll 1$, the individual photons are largely uncorrelated; thus their arrival times must obey Poisson statistics. The second term in Eq. (14.64) gives rise to what is known as *excess* photon noise. It results from the contribution of $\eta\bar{n}$ to the factor $(\eta\bar{n} + 1)$ in Eq. (14.63), and thus it is dominant in the limit $\eta\bar{n} \gg 1$. In this limit the photons are highly correlated and display bunching of the sort associated with Bose–Einstein statistics. This is also the limit where, by the correspondence principle, an entirely classical description of the noise is possible. This noise is due to the intensity fluctuations that are present in any thermal source. This noise is present even in the detection of a thermal source with $\bar{n} \ll 1$, but it is largely masked by the presence of shot noise. The noise associated with the second term in Eq. (14.64) is not present in the detection of a source of constant intensity, such as a well-stabilized laser.

14.4 DETECTION LIMITATIONS IMPOSED BY PHOTON NOISE

Sections 14.1 through 14.3 discuss the fluctuations that are present in any thermal radiation field. These fluctuations provide the fundamental limitation to the sensitivity of a detection system. These general principles are illustrated in the present section by calculating the background-limited NEP and D^* for detection with thermal and photon detectors using the concept of photon noise discussed in the preceding section.

Thermal Detectors

Let us consider an ideal thermal detector of active area A that is maintained at temperature T_0 and is exposed to radiation from a source of temperature T_1 which subtends half angle θ, as shown in Fig. 14.8. It is assumed that the baffles that restrict the field of view of the detector are totally black and are maintained at temperature T_2. The spectral emissivity of the detector, which is assumed to be independent of angle, is denoted $\varepsilon(\nu)$.

Since the detector is not in thermal equilibrium with the radiation background, the simple treatment of the NEP given in Chapter 13 does not apply, and the NEP must be calculated in terms of the fluctuations in the emitted and absorbed power. The fluctuation in the rate at which power is absorbed by the detector from the source can be obtained by integrating the square of Eq. (14.62) (which assumed a narrow spectral interval) over all

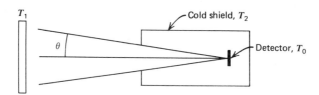

Figure 14.8. Ideal thermal detector of temperature T_0 is exposed to a blackbody background temperature T_1. The field of view is restricted by a shield of temperature T_2.

frequencies, giving

$$\overline{(\Delta P_{abs})^2} = 2\Delta f \int_0^\infty h\nu\varepsilon(\nu)\,\overline{P_\nu}\left[\varepsilon(\nu)\bar{n}_\nu + 1\right]d\nu. \qquad (14.65)$$

In this equation, $\overline{P_\nu}$ denotes the spectral density of the incident power, $\varepsilon(\nu)$ has formally replaced η, and a subscript ν has been added to $\bar{n}$ to illustrate the frequency dependence of $\bar{n}$. Through use of Eq. (14.54) for $\overline{P_\nu}$ with $\Omega_{proj} = \pi \sin^2\theta$, and through use of Eq. (14.51) for $\bar{n}$, this equation can be expressed as

$$\overline{(\Delta P_{abs})^2} = \frac{4\pi \sin^2\theta\,\Delta f\,h^2}{c^2} \int_0^\infty \frac{\varepsilon(\nu)\nu^4\left(e^{h\nu/kT_1} + \varepsilon(\nu) - 1\right)d\nu}{\left(e^{h\nu/kT_1} - 1\right)^2}. \qquad (14.66)$$

In many cases of practical interest, a thermal detector and its housing are cooled so that fluctuations in only the power incident from the source influence the detector performance. In this case, the detector NEP for radiation of frequency ν_s can be obtained from Eq. (14.66) through the relation

$$P_N = \frac{\left[\overline{(\Delta P_{abs})^2}\right]^{1/2}}{\varepsilon(\nu_s)}. \qquad (14.67)$$

One important special case for which the integral in Eq. (14.66) can be evaluated in closed form is that of a black detector [i.e., one for which $\varepsilon(\nu) = 1$]. For this case $\overline{(\Delta P_{abs})^2}$ is given by

$$\overline{(\Delta P_{abs})^2} = 8A \sin^2\theta\sigma kT_1^5 \Delta f, \qquad (14.68)$$

where σ denotes the Stefan–Boltzmann constant defined in Eq. (3.68). [For details on how to evaluate this integral, see P. B. Fellgett, *J. Opt. Soc. Am.*, **39**, 970 (1949) or R. C. Jones, in *Advances in Electronics* V, Academic, New York, 1953. It is interesting to note that had the factor of $(\varepsilon(\nu)\bar{n}_\nu + 1)$ in Eq. (14.65) been replaced by unity, which is equivalent to assuming that the photon arrival times at the detector are truly random, the factor of 8 in this result would have been replaced by a factor of 7.66. Thus the tendency of photons to bunch increases the mean-square power fluctuation by only approximately 5% in this case].

If the detector element and the baffle that defines the field stop are at a finite temperature, they will emit thermal radiation, and resulting fluctuations in the heat flux to the detector must also be taken into account. The exact nature of these fluctuations is very difficult to calculate unless $\varepsilon(\nu) = 1$, and thus this condition will be assumed in the following discussion. [Fluctuations in the thermal radiation emitted by a non-black emitter have been discussed by R. E. Burgess, *J. Phys. Chem. Solids*, **22**, 317 (1961)]. The fluctuation in the rate at which power emitted by the baffle is absorbed by the detector is given by an equation analogous to Eq. (14.68), but with T_2 replacing T_1 and with Ω_{proj} given by $\pi(1 - \sin^2\theta)$, giving

$$\overline{(\Delta P_{abs})^2} = 8A(1 - \sin^2\theta)\sigma k T_2^5 \, \Delta f. \qquad (14.69)$$

In addition, the rate at which thermal radiation is emitted by the detector element is a fluctuating quantity. This fluctuation is also given by an equation analogous to Eq. (14.68), with $\Omega_{proj} = \pi$, since the detector can emit into a full hemisphere, giving

$$\overline{(\Delta P_{em})^2} = 8A\sigma k T_0^5 \, \Delta f. \qquad (14.70)$$

Since the fluctuations in the emission rate and the absorption rates are uncorrelated, the total power fluctuation is given by the sum of Eqs. (14.68), (14.69), and (14.70), giving

$$\overline{(\Delta P_{tot})^2} = 8A\sigma k \left[T_0^5 + T_1^5 \sin^2\theta + T_2^5(1 - \sin^2\theta) \right] \Delta f. \qquad (14.71)$$

The NEP is then given by

$$P_N = \left[\overline{(\Delta P_{tot})^2} \right]^{1/2}, \qquad (14.72)$$

and the specific detectivity by

$$D^* = \frac{\sqrt{A \, \Delta f}}{P_N}$$

$$= (8\sigma k)^{-1} \left[T_0^5 + T_1^5 \sin^2\theta + T_2^5(1 - \sin^2\theta) \right]^{-1/2}. \qquad (14.73)$$

For the special case in which $T_0 = T_1 = T_2$, these expressions for D^* and the NEP agree with those of Eq. (13.18) derived under the conditions of thermal equilibrium.

Photon Detectors

Let us consider an ideal photon detector of area A whose quantum efficiency for radiation of frequency ν is denoted $\eta(\nu)$. A weak signal of optical frequency ν_S and power P_S falls onto the detector. In addition, background power from a blackbody source of temperature T and of half angle θ falls onto the detector. The analysis of this situation differs from that of a thermal detector in several important ways: (1) Fluctuations in the rate at which photoevents occur, rather than fluctuations in the absorbed power, must be considered; (2) fluctuations in thermally *emitted* radiation need not be considered; and (3) photon detectors are generally characterized by a low-frequency cutoff. The mean-square fluctuation in the rate at which photoevents occur is obtained by dividing the square of expression (14.62) for the fluctuation in absorbed power by $h^2\nu^2$ and by integrating this result over all frequencies, giving

$$\overline{(\Delta r)^2} = 2\,\Delta f \int_0^\infty \eta(\nu)\frac{\overline{P_\nu}}{h\nu}\big[\eta(\nu)\bar{n}_\nu + 1\big]\,d\nu. \qquad (14.74)$$

Through use of Eq. (14.54) for $\overline{P_\nu}$ with $\Omega_{\text{proj}} = \pi\sin^2\theta$, and through use of Eq. (14.51) for $\bar{n}_\nu$, this equation can be expressed as

$$\overline{(\Delta r)^2} = \frac{4\pi A \sin^2\theta\,\Delta f}{c^2}\int_0^\infty \frac{\eta(\nu)\nu^2\big[e^{h\nu/kT} + \eta(\nu) - 1\big]\,d\nu}{\big(e^{h\nu/kT} - 1\big)^2}. \qquad (14.75)$$

The background-noise-limited NEP is obtained by equating $(\Delta r)_{\text{rms}} = [\overline{(\Delta r)^2}]^{1/2}$ with the rate $\eta(\nu_S)P_N/h\nu_S$ at which photoevents are produced by the signal radiation. The result is conveniently expressed in terms of the specific detectivity

$$D^* = \frac{\sqrt{A\,\Delta f}}{P_N}$$

$$= \left[\frac{4\pi \sin^2\theta h^2\nu_S^2}{\eta^2(\nu_S)c^2}\int_0^\infty \frac{\eta(\nu)\nu^2\big[e^{h\nu/kT} + \eta(\nu) - 1\big]\,d\nu}{\big(e^{h\nu/kT} - 1\big)^2}\right]^{-1/2}$$

$$(14.76)$$

One special case of this general result entails the assumption that

$$\eta(\nu) = \begin{cases} 0 & \text{for } \nu < \nu_c \\ 1 & \text{for } \nu \geqslant \nu_c, \end{cases} \qquad (14.77)$$

where ν_c denotes the cutoff frequency of the detector. It is often further assumed that the spectral response of the detector is matched to that of the signal to be detected, so that $\nu_S = \nu_c$. Under these assumptions, Eq. (14.76) reduces to

$$D^* = \left[\frac{4\pi \sin^2\theta h^2 \nu_S^2}{c^2} \int_{\nu_S}^{\infty} \frac{\nu^2 e^{h\nu/kT} \, d\nu}{\left(e^{h\nu/kT} - 1\right)^2} \right]^{-1/2}. \tag{14.78}$$

This equation is displayed graphically in Fig. 14.9.

Although Eq. (14.78) assumes rather specialized conditions [i.e., background radiation due to a blackbody source, and a detector whose quantum efficiency given by Eq. (14.78)], it provides a useful standard against which to compare the performance of actual detectors. Many commercial manufacturers of infrared detectors compare the performance of their detectors against the predictions given by Eq. (14.78). One such example is shown in Fig. 14.10.

Some intuition into the nature of photon noise can be obtained through a consideration of a perfect photon detector, that is, one for which $\eta(\nu) = 1$ for all frequencies ν. For this case, the integral appearing in Eq. (14.75) can be evaluated in terms of the Riemann zeta-function [see P. B. Fellgett, *J.*

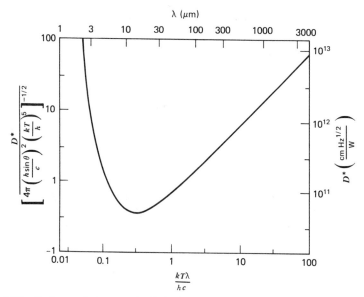

Figure 14.9. Background-photon-noise-limited D^* of an ideal photon detector, from Eq. (14.78). The left-hand and lower axes give the functional form of this equation, while the right-hand and upper axes refer to the particular case of a hemispherical field of view ($\theta = 90°$) and a blackbody background temperature of 300 K.

Opt. Soc. Am., **39**, 970 (1949) for details], and the mean-square fluctuation in the photoevent rate is given by

$$\overline{(\Delta r)^2} = \frac{4\pi A \sin^2\theta \, \Delta f}{c^2} \int_0^\infty \frac{\nu^2 e^{h\nu/kT} \, d\nu}{\left(e^{h\nu/kT} - 1\right)^2}$$

$$= \frac{4\pi^3 A \sin^2\theta \, \Delta f}{3c^2} \left(\frac{kT}{h}\right)^3. \tag{14.79}$$

In order to interpret this result, we note that the mean rate at which photoevents occur is given by

$$\bar{r} = \int_0^\infty \frac{\overline{P}_\nu}{h\nu} \, d\nu$$

$$= \frac{2\pi A \sin^2\theta}{c^2} \int_0^\infty \frac{\nu^2 \, d\nu}{e^{h\nu/kT} - 1}, \tag{14.80}$$

using the same substitutions as in Eq. (14.74). The integral appearing in this equation can also be expressed in terms of the Riemann zeta-function $\zeta(s)$ to yield

$$\bar{r} = \frac{4\pi A \sin^2\theta}{c^2} \left(\frac{kT}{h}\right)^3 \zeta(3), \tag{14.81}$$

where $\zeta(3) \simeq 1.202$. Using Eqs. (14.79) and (14.81), the mean-square fluctuation in r can be expressed as

$$\overline{(\Delta r)^2} = 2.74 \bar{r} \Delta f. \tag{14.82}$$

If photons obeyed classical rather than Bose–Einstein statistics, the factor $[\eta(\nu)\bar{n}_\nu + 1]$ would be absent from Eq. (14.74), which in the limit $\eta(\nu) = 1$ would then be given as

$$\overline{(\Delta r)^2} = 2\,\Delta f \int_0^\infty \frac{\overline{P}_\nu}{h\nu} \, d\nu = 2\bar{r} \Delta f. \tag{14.83}$$

This equation thus constitutes a form of the shot-noise formula (8.42) applied to photons, and it assumes a Poissonian photoevent distribution. The actual result given by Eq. (14.82) is ~ 1.37 times larger than that given by Eq. (14.83), illustrating the increased noise resulting from the tendency of photons to bunch together in phase space.

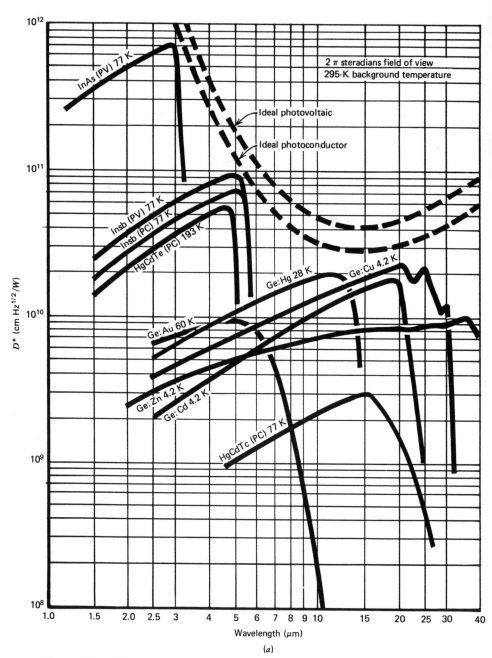

Figure 14.10. Values of D^* for the detection of monochromatic radiation plotted versus the wavelength of the radiation for several commercially available photoconductive (PC) and photovoltaic (PV) detectors. The detector is subjected to a background from a 295-K blackbody subtending a hemispherical field of view. Also shown are theoretical curves for the background-limited D^* for a detector whose cutoff wavelength is equal to the signal wavelength. The curve for a photoconductor is $\sqrt{2}$ times lower than that for an ideal photon detector due to the presence of gr noise. Courtesy of Santa Barbara Research Center.

248

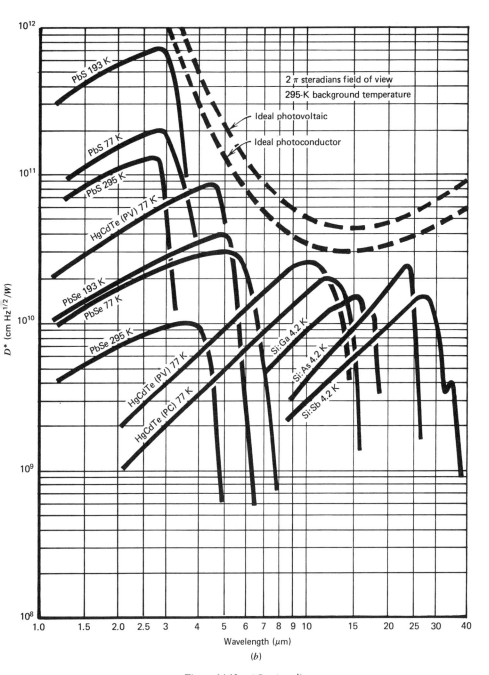

Figure 14.10. (*Continued*)

249

BIBLIOGRAPHY

R. Hanbury Brown and R. Q. Twiss, *Nature*, **177**, 27 (1956); *Proc. R. Soc. London Ser. A*, **242**, 300 (1957); **243**, 291 (1957).

M. V. Klein, *Optics*, Wiley, New York, 1970, Chapter 6.

P. W. Kruse, L. D. McGlauchlin, and R. B. McQuistan, *Infrared Technology*, Wiley, New York, 1962.

R. Loudon, *The Quantum Theory of Light*, Oxford University, London, 1973.

L. Mandel, *Proc. Phys. Soc. London*, **72**, 1037 (1958); **74**, 346 (1959).

L. Mandel and E. Wolf, "Coherence Properties of Optical Fields," *Rev. Mod. Phys.*, **37**, 231 (1965).

E. M. Purcell, *Nature*, **178**, 1449 (1956).

B. Saleh, *Photoelectron Statistics*, Springer, Berlin, 1978.

R. A. Smith, F. E. Jones, and R. P. Chasmar, *The Detection and Measurement of Infrared Radiation*, Oxford University, London, 1968.

K. M. van Vliet, *Appl. Opt.*, **6**, 1145 (1967).

PROBLEMS

1 Calculate the first- and second-order field correlation functions and the spectral density of the power fluctuations for a quasi-monochromatic beam of light whose spectral density has the form of a Gaussian distribution.

2 An object of emissivity $\varepsilon(\nu)$ is in thermal equilibrium with a large isothermal enclosure that surrounds it. Derive an expression for the mean-square fluctuation (within the frequency interval Δf) in the power emitted by the object by making use of Eq. (14.65) for the fluctuation in the absorbed power and by requiring that the blackbody field not be disturbed by the presence of the object.

3 Derive an expression for the photon-noise-limited D^* of a photoconductive detector described by the simple model (which assumed a constant value of the mean carrier lifetime) presented in Sections 10.1 and 10.2. In particular, show how gr noise is influenced by excess photon noise.

4 Calculate the value of D^* for an ideal thermal detector operating at a temperature of 4.2 K. Assume that the detector is a blackbody and is in thermal equilibrium at its operating temperature.

   Now assume that the detector is exposed to background radiation from a 300-K blackbody subtending a 60° (full angle) field of view. What is the value of D^*?

Index